Communications for Survival and Self-Reliance

Michael Chesbro

PALADIN PRESS • BOULDER, COLORADO

Other Books by Michael Chesbro:

The Complete Guide to E-Security:
Using the Internet and E-Mail without Losing Your Privacy

Freeware Encryption and Security Programs:
Protecting Your Computer and Your Privacy

Privacy for Sale: How Big Brother and Others
Are Selling Your Private Secrets for Profit

Privacy Handbook: Proven Countermeasures for Combating
Threats to Privacy, Security, and Personal Freedom

Wilderness Evasion: A Guide to Hiding Out and
Eluding Pursuit in Remote Areas

Communications for Survival and Self-Reliance
by Michael Chesbro

ISBN 1-58160-411-4
Printed in the United States of America

Published by Paladin Press, a division of
Paladin Enterprises, Inc.
Gunbarrel Tech Center
7077 Winchester Circle
Boulder, Colorado 80301, USA
+1.303.443.7250

Direct inquiries and/or orders to the above address.

Visit our Web site at: www.paladin-press.com

TABLE OF CONTENTS

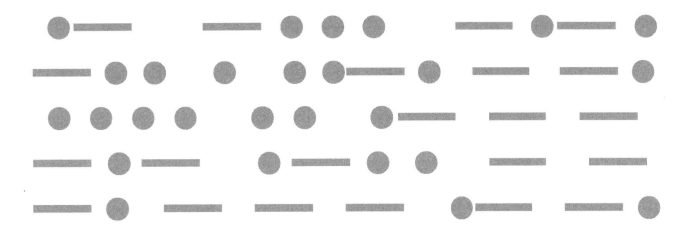

Introduction • 1

CHAPTER 1: AM/FM Broadcast Radio and Shortwave • 7

CHAPTER 2: Licensed Radio Services • 17

CHAPTER 3: Unlicensed Radio Services • 37

CHAPTER 4: Operating Procedures • 45

CHAPTER 5: Digital Communications • 59

CHAPTER 6: Field and Mobile Communications • 69

CHAPTER 7: Power Supplies • 83

CHAPTER 8: Antennas • 87

CHAPTER 9: Codes and Ciphers • 95

CHAPTER 10: Training and Study Options • 113

CHAPTER 11: Radio Retailers and Resources • 115

CHAPTER 12: Suggested Reading • 117

Appendix I: Brevity List • 119

Appendix II: Ops-Code • 127

Appendix III: Amateur Radio Technician Exam Question Pool • 149

For Joni, Skye, Tommi, and Max

A Word About the Law

When you transmit on a radio, your message is sent across the airwaves and is accessible to everyone capable of receiving on the frequency and in the mode you are transmitting. It is important to remember that the airwaves are a shared and finite resource. There are only so many available frequencies. When you transmit, your message may be received miles away, hundreds of miles away, or thousands of miles away in some other country.

Because of the shared and international aspect of radio, operation on many frequencies requires that operators and their stations be properly licensed. In the United States, the FCC controls licensing of radio stations and radio operators.

The FCC establishes rules and regulations for the operation of radio stations. Contrary to regulations promulgated by many government agencies, the regulations that are established by the FCC governing noncommercial radio operations generally make sense.

When operating your private radio station, courtesy is the number-one standard. Be polite, remember that bandwidth and frequencies are shared, and don't do things that intentionally cause harmful interference to others. Courtesy is essential whether you are operating an Amateur Radio station, communicating worldwide, communicating on a CB radio while driving, or a using an FRS radio to maintain contact while shopping at the mall with your family.

Secondly, *comply with the appropriate licensing requirements.* FCC licensing for Amateur Radio requires passing an exam to demonstrate a basic understanding of radio operating procedure and FCC rules and regulations. Other radio services, such as GMRS, require a station license, but there is no examination of any type required. An appropriately licensed station draws little interest or comment. You are just one more station among thousands of others. On the other hand, if you operate without an appropriate license and call sign, you will attract attention and will quickly be identified and reported.

In this book, we discuss certain activities that may violate FCC regulations (or similar regulations in other countries) yet may in fact be perfectly legal in certain jurisdictions. For example, Freeband radio (operating outside of FCC-assigned frequencies on CB radio) is prohibited in the United States, yet some of these same "freeband" frequencies are legally usable in Great Britain. Discussion of a particular communications capability is not a recommendation that one use that capability if it violates the law where one is operating. *It is important to comply with the appropriate rules and regulations governing the communications equipment and techniques you are using.* It is up to you, the reader, to determine the suitability and applicability of the information contained in this book as it relates to your intended use.

A Special Acknowledgment

This book would not have been possible without the constant encouragement, assistance, and friendship of Ray Tougas (AKA: Radio Ray/W7ASA). Ray is a professional engineer, a wilderness survival and preparedness instructor, and an absolute expert in radio communications technology.

Ray has spent months making radio contacts with me, using the various communications modes and methods described in this book. He has introduced me to new communications techniques, helped me test theories and radio techniques, and patiently worked with me to perfect my Morse code ability.

Ray's efforts contributed significantly to the completion of this book. While I acknowledging Ray's efforts and contribution to this work, any errors or omissions are my fault alone.

Thanks, Ray and 73s.

INTRODUCTION

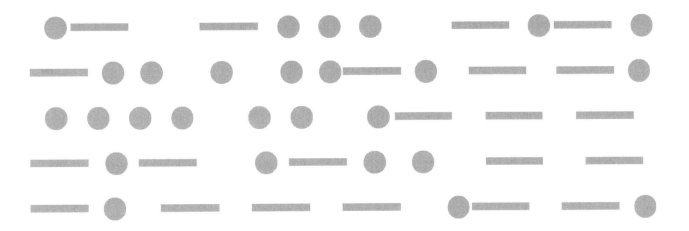

Effective communications are an essential aspect of survival and self-reliance. In recent years there have been numerous books written about survival and self-reliance. In the late 1990s, with the looming specter of Y2K and widespread failure of the infrastructure, it seems that everyone was a Y2K expert and wrote a book to discuss their ideas concerning preparedness. I confess to having read the majority of these books. While most provided excellent information in one form or another, the one aspect conspicuously absent from almost all of these books was a discussion of effective communications for survival and self-reliance.

However, it doesn't take an event like the predicted widespread infrastructure failure of Y2K to make survival communications a valuable asset. A natural disaster or a man-made atrocity can make the need for effective personal communication just as important.

At 5:46 A.M. on the morning of January 17, 1995, a magnitude 6.9 earthquake rocked Kobe, Japan. The shaking lasted about 20 seconds. Electrical power failed, the telephone system was destroyed, and more than 50 fires broke out as a result of the earthquake.

As the sun rose that January morning, rescue personnel were shocked at the amount of destruction. Entire neighborhoods were destroyed. Water mains were broken, leaving the residents without water, causing flooding in some areas, and severely hampering fire-fighting efforts in many others.

Precious hours passed with little response from Tokyo. The earthquake had caused an almost complete failure of the communications infrastructure. The telephone system was destroyed, cellular telephone towers were down, and radio repeater systems were destroyed or quickly failing due to loss of power and depletion of their backup power supplies.

However, Amateur Radio operators were able to report conditions and provide emergency communications until rescue services could restore the failed communications infrastructure.

On September 11, 2001, terrorists attacked the United States of America, destroying the World Trade Center in New York City, damaging the Pentagon, and killing thousands. As we realized that we were under attack and the media began to broadcast reports, I did what thousands of others with friends in New York City were doing at that moment: I picked up the telephone to call my friend and make sure he was OK. I dialed his home number, but all circuits were busy—my call did not go through. Next I called his cellular telephone number, but the cellular towers in that part of the city had just been destroyed. There was no way for my friend to receive or make calls on his cellular telephone. was my friend hurt? Had he been killed?

Unlike many who rely on telephones as their primary and perhaps only means of long-distance rapid communication, my friend and I are radio operators, and we have an established communications plan. I turned on my HAM radio and tuned to the frequency that my friend in New York and I had long ago agreed we would use if one or the other of us was in trouble and we could not make contact by telephone or e-mail.

I began calling my friend, tapping out his call sign in Morse code. Listen, was he there? Nothing, the frequency seemed dead. Again I began sending his call sign . . . and then I heard the clear tones of Morse code in reply: didahdah dit – didah didahdit dit – dahdahdah dahdidah. "We are OK." My friend and his family had not been killed or injured in the attack. They were all OK. They were together, and they were safe.

On a less serious note, I live in an area that does not have cellular telephone coverage. My home is sufficiently remote so that I am beyond the range of the cellular telephone sites. It also seems that my landline telephone service fails a couple times per year. While not a serious problem, such conditions, one might expect, would leave me without outside communication. However, with my radio system I have voice communications, I have data communications, and I can even send e-mail from my radio to any e-mail address on the Internet.

In this book we will discuss communications for survival and self-reliance. This is much more than sending an SOS/MAYDAY call from a sinking ship or using your citizen band (CB) radio to call for road service when your automobile breaks down. We will look at these things, certainly, but also we will learn how to establish a communications system that is independent of outside support, a system controlled by each individual operator.

WHAT IS COMMUNICATIONS FOR SURVIVAL AND SELF-RELIANCE?

When most people think of modern electronic communication, the first thing that comes to mind is the telephone. Ask these people to consider communications for survival and self-reliance, and cellular and satellite telephones come quickly to mind, along with perhaps a CB radio out in the family car. While these are all certainly valid considerations, there is much more involved in communications than these limited options.

Communication for survival and self-reliance must meet two basic criteria to be most effective: (1) the communications system must function effectively in remote areas and under adverse conditions (i.e., no commercial power); and (2) the communications system must be under complete control of the operator.

When we look at the cellular telephone we see that it meets neither of these basic criteria. It is important to remember that your cellular telephone is simply a radio that talks back and forth to a cellular site (or a series of sites if you are moving). If you are in a remote area, where there are no cell-towers/cell-sites allowing the phone to connect to the cellular system, your cell phone will not work. You will receive an "out of range" or "no service" message.

Furthermore, you, the cellular telephone user, do not control the communications system itself. The cellular company from which you purchase your cellular service is in complete control. It is, of course, in the best interest of the company in question to ensure that cellular telephones function properly and calls go through when customers make them, but in the event of some type of widespread emergency or national disaster, cellular systems may not be available for general use. An article in the December 12, 2001, edition of the *New York Times* ("U.S. Considers Restricting Cellphone Use in Disasters") reported that the major cellular companies and the federal government are working to restrict cellular telephone communication, making it available only to government officials during times of national emergency. Such a program, which would give government personnel priority service on all cellular systems during disasters, could be implemented by simply requiring an access code be entered on the telephone keypad in addition to the telephone number before the cellular system would process the call. Unless you were privy to these codes, you would not have access to cellular communication until the system was returned to normal operation.

Even when there is no emergency, cellular systems can become overloaded during times of peak usage. Try to make a telephone call on Christmas morning, or perhaps on Mother's Day, and you may hear: "All circuits are busy now. Please try your call again later." There is nothing wrong with the cellular system; it's just that more people are trying to talk with family and friends than the system can handle.

During times of a widespread emergency, the cellular system may likewise become overloaded. This can have the same, and likely longer-lasting, effect of all circuits being busy. The same emergency situation that causes telephone circuits to become overloaded can also directly affect the cellular system by damaging the system itself. An earthquake, hurricane, or terrorist attack that destroys cellular towers or cellular routing systems can completely disrupt cellular service in any given area for extended periods of time. Even if the cellular sites themselves are not destroyed, they will begin to fail within several hours if commercial power is disrupted. Cellular sites may have battery backup and even emergency generators, but as the batteries are drained and the fuel in the generators is used up, the systems will shut down.

The average handheld satellite telephone looks and operates much like one of the larger cellular telephones. Although satellite telephones suffer some of the same disadvantages as the standard cellular telephone, they have the distinct advantage of working wherever the phone can "see" the sky. This allows the satellite telephone to work in areas without cellular towers or landline telephone service.

Traditionally, satellite telephone usage charges have been fairly expensive, although in the past year or so these prices have come down dramatically and are becoming comparable to standard cellular telephone contracts. One company providing satellite telephone service is Global Star (www.globalstar.com). It offers several different price plans to meet most any need. At the time of this writing, Global Star offered a monthly contract for $49.95 per month, which included 120 minutes of talk time, with additional minutes billed at $0.75 per minute. For someone needing more talk time, the next step up was $99.95 per month, which provided 400 minutes of talk time, with additional minutes billed at $0.65 per minute. Although not cheap, these prices are certainly within the range of most people needing telephone communication from remote areas. However, these rates only apply to home calling originating from within the United States or the Caribbean. If you were to travel to Europe (e.g., Germany) and wanted to use your satellite telephone to call back to the United States, the call would cost you approximately $2.19 per minute, and to receive a call on your satellite telephone while visiting Germany would cost you about $2.47 per minute. These "roaming" rates are about average for anywhere outside of the United States. A prepaid roaming service is available from anywhere in the world to anywhere in the world for $3.69 per minute.

When we look at CB radio, we see that it meets the criteria of operating in remote areas and under adverse conditions. It is also under the complete control of the operator. However, we must ask ourselves whether it will be effective—will it meet our communication needs? In some cases it will, but because CB radio is an almost completely unregulated communications means, we often find that it becomes useless because of harmful interference caused by criminal misuse of the allocated frequencies.

So . . . do cellular and satellite telephones and CB radio have a place in communications for survival and self-reliance? Certainly they do; however, they each suffer from certain limitations and thus should never serve as our only means of communication. Throughout this book we will look at a broad spectrum of communications systems and ways to integrate these various systems into a highly effective personal communications network for survival and self-reliance.

REQUIREMENTS FOR EFFECTIVE COMMUNICATION: SENDER, RECEIVER, AND METHOD

For any type of communication to be effective we must have someone to send a message to, someone to receive that message, and an effective way to carry the message from the sender to the receiver. If any one of these elements is missing, communication will fail.

This may sound like simple common sense as you sit reading this book, but it is a matter that is frequently overlooked. As an example of this, I note that various companies make an "emergency CB radio," packed in a sturdy case and intended to be kept in your vehicle. The advertised intent is that if your vehicle breaks down or you are involved in a minor accident, you can set up this "emergency CB radio" and use it to call for help. We clearly have a sender (you) and a means of communication (the radio), but there is not an established receiver in this equation. Now I know many of you are saying, "There are millions of people with CB radios. Surely if I set up my CB radio and call for help, someone will hear me." There are certainly numerous CB radio operators around, and there is even a reasonable chance that one of these people will hear your call for help. If you are lucky, the person who hears your call for help will respond and send a tow-truck, the police, or an ambulance as needed. On the other hand, you may find that no one responds to your call for help or that your call is seen as a prank.

If you add a designated receiver to this scenario, you have a much better chance of receiving assistance should you need it. If you can arrange with a friend to monitor a given CB channel while you are out, you now have someone actively listening for your call, and prepared to respond.

If we replace the CB radio in our scenario with a 2-meter band HAM radio we have increased the range of our communications considerably, and if we add a high frequency (HF) radio we can conceivably have worldwide communication capability.

PACE

When planning communications, and once we have determined that we have a sender, a receiver, and a method of communication, we add the mnemonic PACE to our plan. PACE stands for **P**rimary, **A**lternate, **C**ontingency, and **E**mergency. PACE planning is a way of building redundancy into our communications plan, thereby ensuring that we will have successful communication when we need it.

For an example of PACE, let's assume that we are using 2-meter band HAM radios in an area with an active repeater system. (NOTE: A repeater is a radio system that receives a radio signal and then retransmits, or repeats, that signal, thereby extending the communications range of that particular transmission. Repeaters may be either *simplex*, which record and play back (or store and forward) the transmission on the same frequency, or duplex, which receive a transmission on one frequency and simultaneously retransmit it on a different frequency. Repeaters are discussed in greater detail in Chapter 3.)

Primary—Repeater # 1 (Frequency) calling at the top of the hour from five minutes before the hour until five minutes after the hour. In this case we plan to establish communications at the top of each hour (or every other hour, every six hours, etc.), using a given repeater in the area as our primary means of communication. This is a reasonable communications plan, but even the best plans fail from time to time. It may be that there are numerous people using the primary repeater at the time we have planned to establish communications and we are thus unable to get our message through, or maybe the repeater we chose for our primary communications channel is off the air for a few hours for regular repair and maintenance. In this case, we simply switch to our alternate communications plan.

Alternate—Repeater # 2 (Frequency) calling from quarter past the hour to 25 minutes past the hour. The alternate communications plan changes repeaters (and thus frequencies) and makes a time adjustment to allow the primary means to be attempted first, but for all practical purposes it mirrors the primary communications plan. However, it may be that something has occurred that has caused both repeaters in our plan to fail. In this case we switch to our contingency plan.

Contingency—Simplex (nonrepeater) Frequency (likely in one of the HF bands) calling from half past the hour until quarter of the hour (e.g., 0930 – 0945 hours). In this case, both the primary and alternate means of communication have failed. Perhaps a storm has damaged both repeaters. In this case, both the sender and receiver will tune to the contingency frequency and make such antenna and power output adjustments as necessary to establish communications on the designated simplex frequency. However, the contingency portion of our PACE communications plan does not necessarily mean that the primary and alternate means have failed. We can activate contingency communications whenever a specific set of circumstances is met. For example, we may choose to activate our contingency communications plan at any time a NOAA weather hazard alert is transmitted. Perhaps our contingency communications plan includes non-real-time message handling (such as PACTOR), and we use it when we have text to send to another station. The contingency portion of PACE can be used at any time an established set of conditions is met.

Emergency—In case of an emergency, it may be unreasonable to wait for a scheduled communication window to send a message. The situation may be such that the communications must be established now! Staying with our scenario of using 2-meter band radios, in case of emergency, the sender will use his radio to contact either the primary or alternate repeater's auto-patch function and use the radio to make a telephone call to the intended receiver (or perhaps to the local 911/emergency service unit).

The above is just a short example of a PACE communications plan. It assumes that the sender and receiver are HAM radio operators with access to a 2-meter repeater system and HF radios. Your communications capabilities may be different, but the concept remains the same. Simply use your own capabilities to fill in each section of PACE.

Proper communications requires planning, practice, and knowledge of the various communications systems available. In this book we will look at how to properly establish a communications network for the purposes of survival and self-reliance. We will discuss the options for establishing both local and international contacts, different types of equipment, and methods of sending your messages securely.

We have looked at the three major requirements for successful communications—a sender, a receiver, and a method—and we have applied these requirements in the PACE system to ensure redundancy and prevent communications failures. Now let's move on and look at the various licensed and unlicensed radio services, other modes of communication, and specific equipment to bring our communications plan together in a useable and highly effective operational package.

CHAPTER 1

AM/FM Broadcast Radio and Shortwave

We are all familiar with AM and FM broadcast radio and television. One would be hard pressed to find a home without a radio and television these days. While broadcast radio and television can provide entertainment, advertisement, and some degree of news and information, we must ask ourselves whether they are of any value as means of communication for survival and self-reliance.

First let us consider television from the perspective of communications for survival and self-reliance. Television has a fairly short broadcast range, with reception beyond around 50 miles from the transmitting tower being fairly poor. You can of course receive television signals from satellite or cable if these systems are available in your area and are operational. Generally, I believe that television has little if any value in survival and self-reliance communications. Television news broadcasts are based on the sound bite and video clip, brief snippets and partial comments related to the topic at hand. Even "news programs" seldom provide the most accurate and up-to-date information. Time constraints, editorial policy, and the need to maintain high viewer ratings tend to make television news little more than entertainment based reporting. If you simply must have a television as part of your survival and self-reliance communications gear, consider a small, battery-powered television. Generally, however, a simple radio will provide much more useful information.

This leaves us with AM and FM broadcast radio and shortwave. Again, with the former we have advertiser-supported news based on the

Portable television sets.

sound bite, complete with the brief snippets and partial comments related to the topic at hand. However, with broadcast radio this tends to be far less of a problem than with television. Shortwave radio isn't exempt from the advertiser-supported, sound-bite type reporting. Shortwave radio broadcasts internationally and is therefore less advertiser focused. However, the primary advantage I see to shortwave radio is that it gives immediate international perspective on a given topic and direct reporting on topics that may not make our local or national news.

CLEAR AM BROADCAST STATIONS

Clear Station AM broadcast radio comprises the AM "super stations" broadcasting with a minimum of 10,000 watts but generally at the maximum 50,000 watts. These stations can be heard throughout the United States, especially at night, as the AM signals are reflected off the ionosphere.

Some of the clear station frequencies are shared by two (or sometimes three) Class A stations. However, when this is the case, these stations are separated by significant geographical distance (e.g., opposite sides of the country or lower-48 states and Alaska). Depending on the propagation into your location, you will receive one or the other of the stations sharing a given frequency. As propagation changes, so might the station you receive.

It takes no special equipment to receive these clear stations. Your average home AM/FM radio is capable of receiving them.

Clear 640 KFI, Los Angeles, CA / KYUK, Bethel, AK
Clear 650 WSM, Nashville, TN / KENY, Anchorage, AK
Clear 660 WFAN, New York, NY / KFAR, Fairbanks, AK
Clear 670 WMAQ, Chicago, IL / KDLG, Dillingham, AK
Clear 680 KNBR, San Francisco, CA / KBRW, Barrow, AK
Clear 700 WLW, Cincinnati, OH / KBYR, Anchorage, AK
Clear 710 WOR, New York, NY / KIRO, Seattle, WA
Clear 720 WGN, Chicago, IL / KOTZ, Kotzebue, AK
Clear 740 KCBS, San Francisco, CA
Clear 750 WSB, Atlanta, GA / KFQD, Anchorage, AK
Clear 760 WJR, Detroit, MI
Clear 770 WABC, New York, NY
Clear 780 WBBM, Chicago, IL / KNOM, Nome, AK
Clear 810 KGO, San Francisco, CA / WGY, Schenectady, NY
Clear 820 WBAP, Fort Worth, TX / KCBF, Fairbanks, AK
Clear 830 WCCO, Minneapolis, MN
Clear 840 WHAS, Louisville, KY / KABN, Long Island, AK
Clear 850 KOA, Denver, CO / KICY, Nome, AK
Clear 870 WWL, New Orleans, LA / KSKO, McGrath, AK
Clear 880 WCBS, New York, NY
Clear 890 WLS, Chicago, IL / KBBI, Homer, AK
Clear 1000 WLUP, Chicago, IL / KOMO, Seattle, WA
Clear 1010 WINS, New York, NY
Clear 1020 KDKA, Pittsburgh, PA / KAXX, Eagle River, AK
Clear 1030 WBZ, Boston, MA
Clear 1040 WHO, Des Moines, IA
Clear 1050 WEVD, New York, NY
Clear 1060 KYW, Philadelphia, PA
Clear 1070 KNX, Los Angeles, CA
Clear 1080 WTIC, Hartford, CT / KRLD, Dallas, TX / KASH, Anchorage, AK
Clear 1090 KAAY, Little Rock, AR / WBAL, Baltimore, MD
Clear 1100 WTAM, Cleveland, OH
Clear 1110 WBT, Charlotte, NC / KFAB, Omaha, NE
Clear 1120 KMOX, St. Louis, MO
Clear 1130 KWKH, Shreveport, LA / WNEW, New York, NY
Clear 1140 WRVA, Richmond, VA
Clear 1160 KSL, Salt Lake City, UT
Clear 1170 KVOO, Tulsa, OK / WWVA,

Wheeling, WV / KJNP, North Pole, AK
Clear 1180 WHAM, Rochester, NY
Clear 1190 KEX, Portland, OR
Clear 1200 WOAI, San Antonio, TX
Clear 1210 WPHT, Philadelphia, PA
Clear 1220 WKNR, Cleveland, OH
Clear 1500 WTOP, Washington, DC / KSTP, St. Paul, MN
Clear 1510 WLAC, Nashville, TN / KGA, Spokane, WA
Clear 1520 WWKB, Buffalo, NY / KOMA, Oklahoma City, OK
Clear 1530 KFBK, Sacramento, CA: WCKY, Cincinnati, OH
Clear 1540 KXEL, Waterloo, IA
Clear 1560 KNZR, Bakersfield, CA / WQEW, New York, NY

These clear AM broadcast stations offer a wide variety of programming, allowing almost everyone to find a station of interest. While these "super stations" are certainly aware of their national (and perhaps international audience), part of their broadcasts focus on the geographic area in which the station is actually located. You can be in California and tune in WBZ 1030 AM in Boston, or you can be in New York and tune in KOMO 1000 AM in Seattle.

One of the other major advantages of clear station AM broadcasts is that many of these stations have a "talk radio" format, thus allowing national feedback on the topic of the day. Calling in to one of the several talk radio programs on a clear AM station allows you make a comment or express an opinion to a national audience.

The CCRadio – Plus. (Photo courtesy of C. Crane Company.)

While any radio with AM reception capability can be used to listen to clear station AM broadcasts, if you develop a strong interest in these broadcasts you may want to get a radio specially designed for AM reception. Arguably the best radio currently available for reception of long distance AM stations is the CCRadio – Plus.

The CCRadio – Plus, along with many accessories (and other electronic equipment), is available from

C. Crane Company
1001 Main Street
Fortuna, CA 95540-2008
Web site: www.ccrane.com

I personally own a CCRadio – Plus and find it to be amazing. Using an external AM antenna (also available from the C. Crane Company), I frequently tune in to Clear Station AM broadcasts from other states. I often tune in to a Clear Station in a state where national news has given some snippet of information about some event. Listening to commentary from a station located in the area of said happening provides much greater insight than the national news sound bite.

Once you discover the world of medium wave (AM) radio, you may want to join the National Radio Club, Inc., P.O. Box 164, Mannsville, NY 13661-0164. The National Radio Club, Inc. (www.nrcdxas.org) was founded in 1933 and is the oldest such radio club currently in existence.

FM BROADCAST RADIO

FM broadcast stations are focused on a specific area. They are local broadcast stations providing news, information, and entertainment to an audience in one regional area. FM broadcast is the radio most of us are familiar with, and we likely have a favorite station that we listen to at home or while driving back and forth to work. It is FM that provides local news and entertainment, but because of the limited range of FM transmissions, stations are limited to serving specific local areas. Don't expect to listen to your favorite Boston FM radio station while visiting New York City. Still, because of their local nature, FM radio stations can provide specifically focused information.

An excellent resource for locating radio stations by location or format is the Radio Locator Web site (formerly the MIT list of radio stations on the Internet) at www.radio-locator.com/cgi-bin/home.

SHORTWAVE RADIO LISTENING

While clear station AM broadcast allows you to tune in the nation, shortwave broadcast allows you to tune in the world. The main advantage of shortwave broadcasts is that you can listen to international news as it is broadcast from the nation where events are happening. This can provide information that simply isn't presented by the highly edited American news services. Additionally, you can learn what perspective other nations have regarding events happening in the United States (or other countries).

49 meter band 5730 – 6205 Khz
41 meter band 7100 – 7595 Khz
31 meter band 9350 – 10000 Khz
25 meter band 11550 – 12160 Khz
22 meter band 13570 – 13870 Khz
19 meter band 15000 – 15710 Khz
16 meter band 17500 – 17900 Khz
13 meter band 21450 – 21850 Khz

The majority of the shortwave broadcast radio stations operate within the above frequency ranges. However, there are certainly stations operating within the remaining shortwave bands: 11-meter band, 25600 – 26100 Khz; 60-meter band, 4750 – 5060 Khz; 75-meter band, 3900 – 4000 Khz; 90-meter band, 3200 – 3400 Khz; and 120-meter band, 2300 – 2500 Khz.

Shortwave listening in North America is generally the best at night, mainly because during the day, overseas broadcasters are not beaming their signals to North America. You will certainly find stations to listen to during the day, but far more stations are available after local sundown.

Grundig Satellit 800 Millennium.

During the day, I find the 19-meter band to be the best, with the 16-meter band and the 22-meter band also offering stations of interest. From about two hours before local sunset until two hours after local sunset, and again about two hours before local sunrise until two hours after local sunrise, the 25-meter band is active. This is because of gray-line propagation. At night the 31-meter band, the 41-meter band, and the 49-meter band seem to be most active, broadcasting to North America.

Shortwave radio broadcasts are an excellent medium for developing a broad world outlook, as well as for studying foreign cultures and languages. As an example, the German International broadcast station Deutsche Welle does a special daily news broadcast just for learners of the German language. The 10 A.M. newscast from the Deutsche Welle German service is read slowly and articulated clearly just for German learners.

The two main guides to shortwave radio broadcasts and frequencies are *Passport to World Band Radio*," published every September; and the *World Radio TV Handbook*, published every January by UK-based WRTH Publications Ltd.

Frequency Characteristics and Propagation

When considering the various frequencies available for communication, we must ask ourselves whether there is an advantage of one frequency over another.

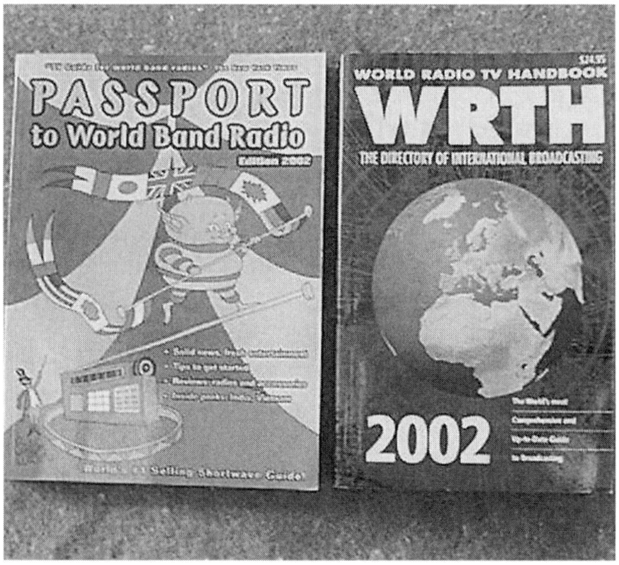

The two main guides to shortwave radio frequencies.

Radio waves travel along the ground and radiate skyward at various angles. These signals travel through space at the speed of light (approx. 186,000 miles per second). Ground wave signals travel near the earth's surface from the transmitter directly to the receiver. Skywave signals travel up from the transmitter (earth) and are refracted off the ionosphere back to the earth. Most short distance communications and all UHF and upper VHF communications are ground wave signals. Most long distance communications are primarily sky wave signals.

The ground wave can be broken into three separate phases. First is the direct wave, which is the signal traveling directly from the transmitting antenna to the receiving antenna. The direct wave is limited to line of sight (plus a small amount of additional distance added by atmospheric refraction and diffraction of the wave around the curvature of the earth.) The distance at which the direct wave can be received is directly affected by the height of the transmitting and receiving antennas. This distance can be increased by increasing the height above ground level of either of the antennas.

The next phase of the ground wave is the ground reflected wave. This is the portion of the ground wave signal that reaches the receiving antenna after being reflected off the surface of the earth. The ground reflected wave can cause signal degradation and even completely cancel out the direct wave signal, if both the direct and ground wave signals reach the receiving antenna at the same time and are 180 degrees out of phase.

The final phase of ground wave propagation is the surface wave. This is the wave that follows the curvature of the earth, allowing the ground wave signal to be received beyond line-of-sight distances. The conductivity and dielectric constant of the earth affect the surface wave.

Sky wave propagation depends on the ionosphere to refract the radio signals back to earth. The ionosphere has four distinct layers: the D layer, E layer, F-1 layer, and F-2 layer.

The D layer is present only during daylight hours and has the primary effect of attenuating radio signals where the transmission path is completely within daylight.

The E layer refracts signals at intermediate distance (less than 1,500 miles) during daylight hours. At night the E layer becomes less intense and is no longer effective as a refracting medium for high-frequency radio signals.

The F layer extends above the earth to a distance of about 240 miles. During the day, sunlight causes ionization of the region, creating two distinct layers (F-1 and F-2). At night the F layers become a single layer at about 170 miles above the earth, and it is this layer that is highly effective in refracting radio signals back to earth via sky wave propagation. The F layer is very useful for long-range radio signals at distances greater than 1,500 miles.

Ionization of the layers of the ionosphere varies on a daily basis based on movements of the earth and the sun and other solar activity. Many of these changes are predictable, since they are based on the earth's rotation, north-south seasonal movement of the sun, 27-day solar rotation, and 11-year sunspot activity cycles.

On the other hand, there are irregular variations in the ionosphere that are not particularly predictable. One of the most common is "sporadic E," resulting from excessive ionization of the E layer, which may block refraction from the F layer, resulting in loss of communication. Sporadic E can also result in the increased refraction of signals of the E layer, increasing communication range by hundreds of miles beyond what is normally expected.

Ionospheric storms lasting from hours to days can disrupt sky wave communication, resulting in flutter and fading of radio signals.

Sudden ionospheric disruption is associated with solar eruptions and results in excessive ionization of the D layer. This results in complete absorption of all signals above approximately 1 MHz, lasting from a few minutes to several hours. When this happens there is a general loss of sky wave signals and all receivers tuned to those signals seem to go dead.

One very effective way to check radio propagation is to listen to the time broadcasts on the various frequencies from different areas of the world. Keep a record of which stations you can hear at what times and on which frequencies. You will soon see a pattern and will then be able to check propagation changes easily by tuning to time stations that you can usually hear.

Time and Frequency Stations

Location	Frequencies (MHz)	Call Sign
Buenos Aires, Argentina	5.000, 10.000, 15.000	LOL
Lyndhurst, Australia	2.500, 5.000, 8.634, 12.984, 16.000	VNG
Ottawa, Ontario, Canada	3.330, 7.335, 14.670	CHU
Xiang, China	5.000, 5.430, 9.351, 10.000, 15.000	BPM
Fort Collins, Colorado	2.500, 5.000, 10.000, 15.000, 20.000	WWV
Kekaha, Kauai, Hawaii	2.500, 5.000, 10.000, 15.000	WWVH
Tokyo, Japan	2.500, 5.000, 8.000, 10.000, 15.000	JJY
Irkutsk, Russia	5.004, 10.004, 15.004	RID
Novosibirsk, Russia	4.996, 9.996, 14.996	RWM

As a generalized overview, we can look at the following frequencies/bands that might be useful for survival and self-reliance communications.

1240 MHz Band (1240.000 MHz – 1300.000 MHz)—The 1240 MHz band is similar to the 900 MHz band. There are a few FM repeaters here. Many Amateur Radio operators conduct experiments in the 1240 MHz band.

900 MHz Band (902.000 MHz – 928.000 MHz)—The 900 MHz band is shared with many other non-amateur services, such as radar, fixed, and mobile operations and scientific and medical industrial services. There are a few Amateur Radio repeaters set up for the 900 MHz band. The 900 MHz band is only an Amateur Radio band in Region 2 (the Americas). Amateur Radio operators in Europe and Asia do not operate on 900 MHz.

GMRS (462.5625 MHz – 467.7250 MHz)—The GMRS frequencies sit just above the Amateur 440 MHz band and have almost identical characteristics.

440 MHz Band (70-Centimeter Band) (420.000 MHz – 450.000 MHz)—The 440 MHz band runs a close second to the 2-meter band as far as use by Amateur Radio operators on the UHF/VHF frequencies.

222 MHz Band (222.000 MHz – 225.000 MHz)—The 222 MHz band is one of the less used but completely acceptable bands for local communication. Sitting between the highly popular 2-meter band and 440 MHz band, the 222 MHz band has similar characteristics but lacks the extensive repeater systems found on 2 meters and 70 centimeters (although there are 222 MHz repeaters).

2-Meter Band (144.000 MHz – 148.000 MHz)—The 2-meter band is the workhorse of the Amateur Radio UHF/VHF bands. This is where most radio operators start after obtaining their technician class Amateur Radio license. There are extensive networks of repeaters established for the 2-meter band. Most Amateur Radio operators have 2-meter capability.

6-Meter Band (50.000 MHz – 54.000 MHz)—The 6-meter band is called the "magic band." The 6-meter band sits between the HF frequencies and the VHF frequencies. When the band is "open," long-range communication is possible, and at these times, there will be a large number of radio operators working these openings. At other times, the band is less used, but there is still significant activity on 6 meters. There are 6-meter repeaters in all states and throughout Canada. The 6-meter band is excellent for local communications and is useable by all Amateur Radio license classes.

10-Meter Band (28.000 MHz – 29.700 MHz)—The 10-meter band is primarily a daylight hours band. It is most effective for long-range contacts during the winter months. During years of peak sunspot activity, worldwide contacts can be made using low-power and simple antennas. When sunspot activity does not allow for DX contacts, 10 meters is a useful band for local communications. Many areas have 10-meter FM repeater systems set up to facilitate this.

11-Meter Band (26.965 MHz – 27.405 MHz)—This is the CB band, and like the 10- and 12-meter bands on either side of it, the 11-meter band is heavily affected by the sunspot cycle.

12-Meter Band (24.890 MHz – 24.990 MHz)—The 12-meter band is heavily influenced by the sunspot cycle. During years of peak sunspot activity, worldwide contacts are common. During years with minimal sunspot activity, the 12-meter band becomes much less active, except for local communications.

15-Meter Band (21.000 MHz – 21.450 MHz)—The 15-meter band is very effective for long-range contacts during years with favorable sunspot activity. Communications on this band are generally most effective during daylight hours and during the winter months.

17-Meter Band (18.068 MHz – 18.168 MHz)—The 17-meter band provides for strong regional signals and good long-range communication. A lot of conversational communication (rag chewing) can be found on 17 meters, as well as a fair amount of mobile activity.

20-Meter Band (14.000 MHz – 14.350 MHz)—The 20-meter band is very effective for long-distance communication. During peak sunspot cycle years, worldwide communication is possible at almost any time of the day or night. Even when sunspot activity does not provide for optimum communication, worldwide contact is still frequently possible with a little extra effort.

30-Meter Band (10.100 MHz – 10.150 MHz)—The 30-meter band is restricted to low-power

(200-watt maximum output) FSK (Frequency Shift Keying) and CW communications only. However, long-distance communications are easily possible on the 30-meter band. Because the band is restricted to FSK/CW communications it tends to be less crowded than the 20- and 40-meter bands on either side of it. Effective communication on 30 meters is possible at any time of the day or night, and the band is somewhat less affected by sunspot activity than its 20-meter neighbor.

40-Meter Band (7.000 MHz – 7.300 MHz)—The 40-meter band is generally "open" throughout the year and at any time of day or night. A large number of both regional and worldwide DX nets are found on 40 meters. Many foreign "broadcast" stations also operate in the 40-meter band and may overpower Amateur Radio communications.

80-Meter Band (3.500 MHz – 4.000 MHz)—The 80-meter band is a very effective regional communications band. Worldwide communications are possible at night. The band is reliable year round, and one will frequently find message traffic nets operating on 80 meters.

160-Meter Band (1.800 MHz – 2.000 MHz)—The 160-meter band sits at the top end of the AM broadcast band. It provides for effective local and regional communications, both during the day and at night. Worldwide communication is problematic but is possible at night during the winter months.

NOAA WEATHER RADIO

NOAA Weather Radio broadcasts continuous weather information on a 24-hour basis. The weather information is obtained directly from the National Weather Service offices throughout the United States. This weather information is updated every one to three hours (or more frequently if needed).

It is usually possible to receive the weather radio broadcast from most anywhere in the United States, even in some fairly remote areas. When you are camping or traveling in some out-of-the-way area, it is very useful to be able to receive current weather reports.

NOAA Weather Radio Frequencies

Channel	Frequency	Channel	Frequency
1	162.550 MHz	6	162.500 MHz
2	162.400 MHz	7	162.525 MHz
3	162.475 MHz	8	161.650 MHz *
4	162.425 MHz	9	161.775 MHz *
5	162.450 MHz	10	163.275 MHz *

* The NOAA Web site states, "Many NOAA Weather Radios are also programmed for three additional frequencies: 161.650 MHz (marine VHF Ch 21B), 161.775 MHz (marine VHF Ch 83B), and 163.275 MHz. The initial two frequencies are used by Canada for marine weather broadcasts. 163.275 MHz was used by the National Weather Service for internal coordination in the event of a power outage but is no longer in active use."

Another major advantage of the NOAA weather radio system is its alert function. One can purchase special radios that are designed to listen for special alert tones from the NOAA and sound an alarm and broadcast severe weather notices from the NOAA. These weather alert radios are very useful to have at home to warn of incoming storms. The NOAA system can also warn of other local emergencies or special conditions that might affect a specific area.

Many radios have the "weather band" built in as an additional function; however, any radio that can be tuned to the NOAA frequencies can receive NOAA broadcasts. Tune your radio to each of the NOAA listed frequencies and find the specific frequency that serves your area, and you will have instant weather information 24 hours per day thereafter.

CHAPTER 2

Licensed Radio Services

The licensed radio services are, as the name would imply, those that require a license from the FCC (in the United States) in order to operate legally. While there is never any license required to listen, it is important to be properly licensed if you intend to transmit.

The primary licensed service related to our purposes of survival and self-reliance is the Amateur Radio Service (also called HAM radio). We will look at frequency allocations for Amateur Radio and at the band plan used by Amateur Radio operators. Within the Amateur Radio Service are emergency service organizations, such as ARES (Amateur Radio Emergency Service) and RACES (Radio Amateur Civil Emergency Service), and there are affiliate services, such as MARS (Military Affiliate Radio System), which we will look at. There are networks of repeaters allowing long-range communication with nothing more than a hand-held radio.

We will also consider the General Mobile Radio Service (GMRS), intended to provide localized personal communications, and we will look at other licensed services, such as Marine Band Radio and Land Mobile Radio, and consider what use these services may be for survival and self-reliance communications.

HAM RADIO

Anyone truly interested in survival communications should take the time to earn his Amateur Radio license. Of all the possibilities for establishing a communications network for survival and self-reliance, HAM radio is the single best choice for most people because it provides the best and most complete means of communication in survival situations or at times when the normal communications infrastructure fails.

Although Amateur Radio stations can be large and complex affairs, they can also be very small, portable setups intended for use in the field. In fact, "Field Day" is a yearly event for Amateur Radio operators where we set up our radio stations and operate without the use of commercial power, often from a remote area, while making contacts around the world.

Once you earn your Amateur Radio license, you will have operating privileges on several different bands and frequencies, with specific privileges depending on the class of license you hold. Your basic Amateur Radio license (technician class) gives you operating privileges on all the UHF/VHF bands from 6 meters through the 1240 band. In these bands you will find numerous activities, repeater systems to allow you to extend your communications range far beyond the capability of your radio alone, networks providing information on a wide range of topics of interest to radio operators, and much, much more.

For survival and self-reliance communications, you will find that you suddenly have local communications capability that you could only wish for using any of the unlicensed services. Using handheld radios (walkie-talkies) operating in the UHF/VHF amateur bands, you can make contacts at distances of several miles. Using the same handheld radio, you can make contacts through an established repeater system at distances of hundreds of miles. Using an auto-patch system, you can use your radio to make a telephone call. You can set up a packet radio system and send text from your home computer over the radio to another Amateur Radio operator. In these bands you also gain access to Amateur Radio satellites. (Yes! You can send signals through a satellite using your HAM radio.) You can also set up your own amateur television station here.

With the most basic Amateur Radio operator's license, you now have operating privileges that give you practically unlimited local and regional communications. It is now simply a matter of choosing your equipment and which of the many activities meet your communication needs.

UHF/VHF Band	**Frequency Range**	**National Calling Frequency**
1240 Band	1240.000 MHz – 1300.000 MHz	1296.100 SSB/1294.500 FM
900 Band	902.000 MHz – 928.000 MHz	902.100 SSB
440 (70 cm) Band	420.000 MHz – 450.000 MHz	432.100 SSB/446.000 FM
222 Band	222.000 MHz – 225.000 MHz	222.100 SSB/23.500 FM
2 Meters	144.000 MHz – 148.000 MHz	144.200 SSB/146.520 FM
6 Meters	50.000 MHz – 54.000 MHz	50.125 SSB/52.525 FM
<u>HF Bands</u>		
10 Meters	28.00 MHz – 29.700 MHz	28.400 SSB
12 Meters	24.890 MHz – 24.990 MHz	
15 Meters	21.000 MHz – 21.450 MHz	
17 Meters	18.068 MHz – 18.168 MHz	
20 Meters	14.000 MHz – 14.350 MHz	
30 Meters	10.100 MHz – 10.150 MHz	
40 Meters	7.000 MHz – 7.300 MHz	
75/80 Meters	3.500 MHz – 4.0000 MHz	
160 Meters	1.800 MHz – 2.000 MHz	

The UHF/VHF bands give you excellent local and regional communication and, under certain conditions, long-distance communication. I have personally communicated from western Washington state to Nova Scotia, Canada, using only 5 watts of power on the 6-meter band. It is important to note, however, that such long-range communications using 6 meters is *not* the norm.

There will almost certainly come a time when you will want to have worldwide communication capability, and this is where the high frequency (HF) bands come into play. Access to the HF bands requires an upgrade of your Amateur Radio license from technician to general class. Thereafter, you are authorized to operate on all of the Amateur Radio bands (and most frequencies within those bands).

High frequency radio is like having your very own shortwave radio station. In fact, you will hear commercial shortwave foreign broadcast stations on some of the Amateur Radio bands (especially on the 40-meter/7 MHz band). Operating on the HF bands, you have worldwide communication capability at your fingertips, and you have complete control over your own station. There is no reliance on repeaters (although there are repeaters available on the 10-meter band). You can communicate with friends and family (assuming they are also HAM radio operators) in other states and in other countries. Furthermore, you will likely make friends on the air all around the world. If you hear a news report about something going on in another country, you may be able to find a HAM radio operator in that country who can give you firsthand comments on that event.

If you are interested in participating in emergency response, Amateur Radio gives you the opportunity to do so, through organizations such as ARES and RACES. Furthermore, many Amateur Radio operators are set up to operate without the need for commercial power. Thus, in times of a natural disaster or other widespread emergency, your HAM radio may provide you with communication when the normal infrastructure fails.

Band Plan

The band plan is an operating plan that helps keep some degree of order to communications within the Amateur Radio bands. Certain portions of each band are designated for specific types of communication. Primarily, this is broken down into phone (voice) and nonvoice types of communications. Furthermore, depending on your Amateur Radio license class, you are restricted to operating in selected portions of each band. You are authorized to operate only on those frequencies designated for your license class. You may also, of course, operate on those frequencies designated for lesser license classes than the one you currently hold.

The following band plan shows operating modes and license class privileges. Although there are currently only three license classes being issued by the FCC (technician, general, and extra), I have included the novice and advanced license class information in the band plan because there are still people who hold novice and advanced licenses and have specific privileges and restrictions under those old license classes.

10 Meters 28.100 MHz – 29.700 MHz
Novice and Technician Plus
28.100 – 28.300 MHz: CW, RTTY/Data
28.300 – 28.500 MHz: CW, Phone
(Maximum power, 200 watts)

General, Advanced, Amateur Extra
28.000 – 28.300 MHz: CW, RTTY/Data
28.300 – 29.700 MHz: CW, Phone, Image

12 Meters 24.980 MHz – 24.990 MHz
General, Advanced, Amateur Extra
24.890 – 24.930 MHz: CW, RTTY/Data
24.930 – 24.990 MHz: CW, Phone, Image

15 Meters 21.100 MHz – 21.450 MHz
Novice and Technician Plus
21.100 – 21.200 MHz: CW Only

General
21.025 – 21.200 MHz: CW, RTTY/Data
21.300 – 21.450 MHz: CW, Phone, Image

Advanced
21.025 – 21.200 MHz: CW, RTTY/Data
21.225 – 21.450 MHz: CW, Phone, Image

Amateur Extra
21.000 – 21.200 MHz: CW, RTTY/Data
21.200 – 21.450 MHz: CW, Phone, Image

17 Meters 18.110 MHz – 18.168 MHz
General, Advanced, Amateur Extra
18.068 – 18.110 MHz: CW, RTTY/Data
18.110 – 18.168 MHz: CW, Phone, Image

20 Meters 14.025 MHz – 14.350 MHz
General
14.025 – 14.150 MHz: CW, RTTY/Data
14.225 – 14.350 MHz: CW, Phone, Image

Advanced
14.025 – 14.150 MHz: CW, RTTY/Data
14 – 175-14.350 MHz: CW, Phone, Image

Amateur Extra
14.000 – 14.150 MHz: CW, RTTY/Data
14.150 – 14.350 MHz: CW, Phone, Image

30 Meters 10.100 MHz – 10.150 MHz
General, Advanced, Amateur Extra
10.100 – 10.150 MHz: CW, RTTY/Data
(Maximum power, 200 watts)

40 Meters 7.100 MHz – 7.300 MHz
Novice and Technician Plus
7.100 – 7.150 MHz; CW Only

General
7.025 – 7.150 MHz: CW, RTTY/Data
7.225 – 7.300 MHz: CW, Phone, Image

Advanced
7.025 – 7.150 MHz: CW, RTTY/Data
7.150 – 7.300 MHz: CW, Phone, Image

Amateur Extra
7.000 – 7.150 MHz: CW, RTTY/Data
7.150 – 7.300 MHz: CW, Phone, Image

80 Meters 3.675 MHz – 4.000 MHz
Novice and Technician Plus
3.675 – 3.725 MHz: CW Only

General
3.525 – 3.750 MHz: CW, RTTY/Data
3.850 – 4.000 MHz: CW, Phone, Image

Advanced
3.525 – 3.750 MHz: CW, RTTY/Data
3.775 – 4.000 MHz: CW, Phone, Image

Amateur Extra
3.500 – 3.750 MHz: CW, RTTY/Data
3.750 – 4.000 MHz: CW, Phone, Image

160 Meters 1.800 MHz – 2.000 MHz
General, Advanced, Amateur Extra
1.800 MHz – 2.000 MHz: CW, Phone, Image, Data

CW (Continuous Wave, or Morse code) is permitted on all frequencies but is exclusive in those frequency ranges indicated.

Data refers to digital modes, including but not limited to AMTOR, CLOVER, PACTOR, and PSK31. (See Chapter 5 for more on digital communications modes.)

License Classes

As of April 15, 2000, the FCC restructured the Amateur Radio licenses classes to include only three: technician, general, and extra. Under this restructuring, there is no Morse code requirement for the technician class license, and the Morse code speed was reduced to five groups per minute for the general and extra class licenses.

There is a fairly simple written exam, covering basic operating procedures, required to earn your technician license. The general license written exam is only slightly more complicated, adding a bit of communications electronics knowledge to the requirements. You must also pass the Morse code test at five groups per minute to earn your general license. The extra license written exam is quite complex, requiring a strong knowledge of communications electronics and radio theory and operating procedures. However, with a bit of dedication and study, the Amateur extra license can be earned too.

Once you pass the exam for your radio license, you will be assigned a call sign in the FCC database within several days after your exam date (usually around 10 days or so). You are then free to begin transmitting on the Amateur Radio frequencies assigned to the license class you have earned. You will receive the actual paper copy of your Amateur Radio license in the mail about a month later. (One note regarding the information on your Amateur Radio license. It is a public record, so your name and address, along with your call sign, can be quickly found by searching the various online call sign databases. If you don't want your physical address listed in these databases, you might want to establish a post office box to use as your address on your Amateur Radio license.)

Exam Question Pool

The Amateur Radio exam question pool is

available to anyone who wants to have it. All of the questions that may be on any Amateur Radio test, along with the correct answers to those questions, are published and available for study.

The current question pool for each exam is published in several places on the Internet and in various books. When studying for your Amateur Radio license, you might want to review the question pool for the license class for which you are studying at the American Radio Relay League (ARRL) Web site: www.arrl.org/arrlvec/pools.html.

A copy of the technician class license pool is included in Appendix III of this book.

You can also obtain a copy of the technician license question pool and associated graphics for $1.50 from ARRL/VEC, 225 Main Street, Newington, CT 06111. Request the "2003 technician class question pool graphics."

Call Sign Databases

The FCC maintains an online database of all call signs and radio station licenses it has issued. This database can be found at http://wireless.fcc.gov/uls/. The database contains only FCC-issued license information but is the only "official" listing of license information online. The FCC database allows one to find amateur, aircraft, commercial/restricted, GMRS, and ship license information. It does not contain information on any foreign-issued radio licenses.

There are other databases that allow one to research foreign radio licenses, as well as containing the Amateur Radio license information from the FCC. Two of the most detailed Amateur Radio license databases are the QRZ.Com database at www.qrz.com/callsign and the Buckmaster Database at www.buck.com/cgi-bin/do_hamcall. Other call sign databases are available through ARRL at www.arrl.org/fcc/fcclook.php3, the University of Arkansas at Little Rock at http://callsign.ualr.edu/callsign.shtml, and various other radio clubs and organizations.

For U.S. Amateur Radio license information (on licenses issued by the FCC), all of the listed databases contain the same information, which is obtained directly from the FCC. When it comes to non-FCC-issued license information, there may be slight variations in the content of these databases. This is because foreign licensees may register their license information with one database but not with another. Additionally, databases such as QRZ and Buckmaster allow license holders to add supplementary information to their call sign listing. This additional information may include a short biography, a photograph, and information about radio-related interests, memberships, and awards.

ARES and RACES

Amateur Radio operators have two organizations—the Amateur Radio Emergency Service and the Radio Amateur Civil Emergency Service—that are specifically focused on providing communications during disasters and under emergency conditions.

ARES is organized and controlled by the ARRL, and RACES is organized and controlled by the FCC. In general, however, ARES and RACES tend to be combined by the groups of radio operators working emergency communications.

Both services are staffed by volunteers. These volunteers work under the coordination of the local civil defense organization, usually the state department of emergency management. ARES/RACES volunteers provide communications support to state and local emergency service agencies when emergency conditions overload the in-place emergency service communications system.

ARES/RACES radio operators are dedicated to operating under emergency conditions and are thus excellent people to contact when developing your own survival communications capabilities. ARES/RACES provide significant information about field communications and operations. Below is the ARES personal equipment checklist for radio operators deploying to participate in field operations. (I have included a few brief additions and comments from my own experience.)

If you are interested in participating in an active emergency communications network after earning your Amateur Radio license, I encourage you to look into joining ARES/RACES.

ARES Personal Equipment Checklist

Forms of Identification	*Mick's Additions/Comments*
ARES Identification Card	
FCC Amateur Radio License	
Driver's License	
Radio Gear	
VHF Transceiver	*Operating Manual for VHF Radio*
Microphone Capability	*Microphone with DTMF Tone*
Headphones	
Power Supply/Extra Batteries	*Amplifier (for 2-meter radio)*
Antennas with Mounts	*Backpacker Quad/Folding J-Pole Antenna*
Spare Fuses	
Patch Cords/Adapters	
(BNC to PL-259/RCA Phone to PL-259)	
SWR Meter	
Extra Coax Communications	*GMRS (FRS) Radio for Local Area*
Writing Gear	
Pen/Pencil/Eraser	
Clipboard	
Message Forms	*Survival Communications Book (This Book)*
Logbook	*Current Year Repeater Directory*
Note Paper	*Signal Operating Instructions/Commo Plan*
Personal Gear (Short Duration)	
Snacks	
Liquid Refreshment	*Thermos with Hot Coffee/Tea, etc.*
Throat Lozenges	
Personal Medication	
Aspirin	
Extra Pair of Prescription Glasses	
Sweater/Jacket	
Personal Gear (72-Hour Duration)	
Foul-Weather Gear	
Three-Day Supply of Drinking Water	
Cooler with Three-Day Supply of Food	
Mess Kit with Cleaning Supplies	*Camp Stove with Fuel for Cooking*
First-Aid Kit	
Personal Medication	
Aspirin	
Throat Lozenges	
Sleeping Bag	*Tent/Tarp/Shelter*
Toilet Articles	
Mechanical or Battery-Powered Alarm Clock	

Flashlight with Batteries/Lantern	
Candles	*Chem-Light/Light Sticks*
Waterproof Matches	
Extra Pair of Prescription Glasses	
<u>Toolbox (72-Hour Duration)</u>	
Screwdrivers	
Pliers	
Socket Wrenches	
Electrical Tape	
12/120-Volt Soldering Iron	*Spare Wire/Clips/Connectors*
Solder	
Volt-Ohm Meter	
<u>Other (72-Hour Duration)</u>	
HF Transceiver	*Operating Manual for HF Radio*
Hatchet/Ax	*Morse Code Key*
Saw	*G5RV Antenna*
Pick	*Antenna Tuner*
Shovel	
Siphon	
Jumper Cables	
Generator/Spare Plugs and Oil	
Kerosene Lights/Camping Lantern or Candles	
3/8-Inch Hemp Rope	*Cord to Pull Antenna into Trees*
Highway Flares	*Signal Flares/Smoke Signals*
Extra Gasoline & Oil	

Repeaters

A repeater is used to extend the range of UHF/VHF (and 10-meter HF) FM radios. A repeater is normally installed at the highest location available, such as the roof of a tall building or a mountaintop. This allows it to transmit signals beyond the range possible from lower locations. Even in fairly open areas the repeater extends the range of communications simply due to the fact that it is retransmitting signals.

A repeater is a duplex device. This means that it receives on one frequency and transmits on a different one. Radios that are intended for use with repeaters have an automatic shift built into them. Thus, when you key the microphone on your radio, the radio shifts to the repeater's input frequency. While you are transmitting, the repeater receives your message on its input frequency and simultaneously retransmits it on the repeater's output frequency. When you release the microphone key, your radio shifts back to the repeater's output frequency, allowing you to hear any reply to your transmission.

Most populated areas of the United States and Canada have repeater systems in place. These are usually operated and maintained by HAM radio clubs, and sometimes by individual HAM radio operators. Even some fairly remote areas might have repeaters in place where HAM radio operators have set them up. However, it should be noted that repeaters do NOT cover all areas of the United States, and there will always be places where there will be no repeater within range of your radio.

A very comprehensive listing of repeaters is published by the American Radio Relay League. The *ARRL Repeater Directory* lists the repeaters, their locations, input and output frequencies, sponsor and call sign, and notes about the repeater (such as CTCSS tones required to access it). Most repeaters operated by HAM radio operators and their associated clubs are open repeaters, available for use by any licensed HAM radio operator. In many cases, repeaters have additional functions associated with them, such as autopatch, allowing one to make local telephone calls through the repeater, and networking, allowing one to link two or more repeaters together for transmission over greatly extended areas. These extra functions are very often freely available to everyone using the repeater, although on occasion they may be restricted to members of the HAM radio club that maintains the repeater. In these cases, there is usually a series of DTMF tones required to access the special functions of the repeater. These access tones are published in club newsletters or repeater guides provided to members.

Repeater Mapping

Once you have an FM radio set up and a copy of the *ARRL Repeater Directory* or other list of repeaters in your area, you should put together a repeater map. To do this, simply begin by attempting to make a contact through every repeater on your list. Once you know that you can reach a repeater from a specific location (e.g., your home), identify the specific location of that repeater and plot the location on a map. List the repeater frequencies, CTCSS tones, and other related information in your notes. As you make contacts through the various repeaters, plot the location of the contact on your map. You will soon have an accurate map of all areas you can reach using your FM radio. This comes in very handy when you have mobile stations traveling in a given area. Just look at your repeater map and choose the repeater giving you the best coverage for the area in which you are traveling.

Simplex Repeaters

While I have said that a repeater is a duplex device (and most are), there is such a thing as a simplex repeater. The simplex repeater transmits and receives on the same frequency. This is accomplished by a record and playback function (also called a store and forward function). The simplex repeater consists basically of a digital recording circuit and a timing circuit. The simplex repeater is connected to a radio's audio output (speaker/earphones connection), allowing the simplex repeater's digital recording circuit to record any signals received by the radio. The simplex repeater also connects through the timing circuit to the radio's audio input line (external microphone connection). When the simplex repeater's digital recording circuit has recorded a transmission, the timing circuit switches to the input line of the radio and replays whatever has been recorded, thereby repeating/retransmitting the signal.

Simplex repeaters are excellent tools for working in remote areas where there are no duplex repeaters in place and for temporary installations where repeater service is needed for a short time and the expense of a duplex repeater simply cannot be justified. A simplex repeater can be built for just a few dollars, if you have electronics skills, or purchased from various sources for around $100. Following are some sources for simplex repeaters:

MFJ-662 Simplex Pocket Repeater
www.mfjenterprises.com/products.php?prodid=MFJ-662

Radio Shack

The Radio Shack Simplex Repeater was discontinued in 2001 but is still available as back stock from some Radio Shack stores and from places such as eBay Online Auctions.

NHRC 3+
www.nhrc.net/nhrc-3plus/index.shtml

Echo Station
www.synergenics.com/sc/

When using a simplex repeater, you will actually hear your transmission played back, since the simplex repeater stores and forwards your signals. This is different from a duplex, where your transmission is repeated on a different frequency as you send it. It is also important to remember to keep your transmissions short when using a simplex repeater. Most of the digital recording circuits used in simplex repeaters have a recording time of 30 – 60 seconds. If your simplex repeater records for only 30 seconds but you talk for 45 seconds before allowing the simplex repeater to replay your message, some of that message will obviously be lost.

It is also important to use a CTCSS or DCS (digitally coded squelch) so that your simplex repeater does not record and retransmit static or transmissions from others using the frequency but not wanting to use the simplex repeater.

A simplex repeater can be used with most any radio that has external speaker and external microphone connections. Because transmit and receive are functions of the radio to which the simplex repeater is connected, the simplex repeater can be used on any band or frequency (although repeaters are most commonly used with UHF/VHF FM systems).

For survival and self-reliance, simplex repeaters are excellent tools. A simplex repeater, radio, antenna, battery pack to power the radio and simplex repeater, and perhaps a solar panel to trickle-charge the battery pack can be assembled relatively inexpensively. The simplex repeater, radio, and battery pack should all be enclosed in a weatherproof container. The antenna and solar panel, of course, are extended outside of the container—the antenna positioned for best reception of signals and the solar panel positioned to collect the greatest amount of sunlight.

These simplex repeaters are highly portable and are easily placed on hilltops, ridge lines, or tall buildings, to quickly extend the communications range of your radio. In their simplest configuration you just plug your simplex repeater in to your radio and you are ready to go.

I have used this "plug-and-play" setup on occasion when I was able to leave the radio/repeater inside a building or other protected location. More frequently, however, I put the radio and repeater in a small plastic (watertight) container and place it in an outdoor location. The plastic container does not degrade the radio's signal, but it protects the radio and repeater from the elements. It can be placed in the branches of a tree, on a hilltop, or at any other good location for receiving and repeating signals.

For home use, I keep a simplex repeater connected to a GMRS base station. When out with GMRS handheld radios, I found that it was usually possible to talk back to the house from anywhere on my property but that handheld

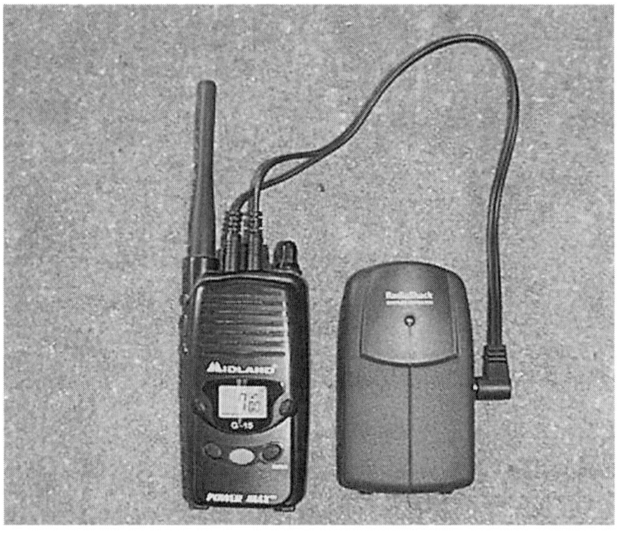

Radio Shack Simplex Repeater connected to a Midland G-15 GMRS Handi-Talkie.

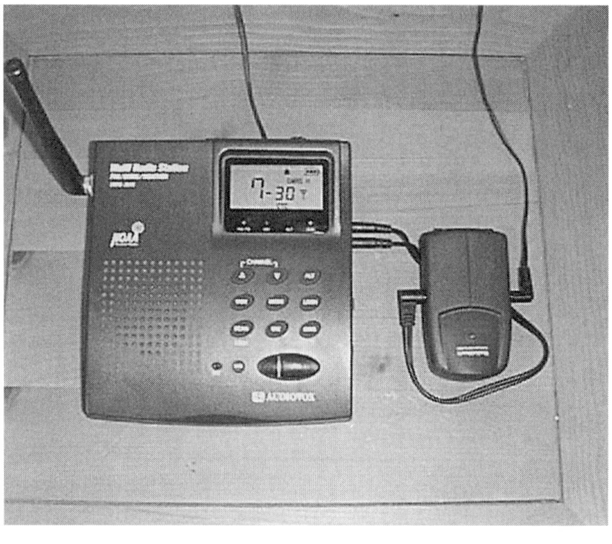

GMRS base station connected to a simplex repeater.

radios couldn't always communicate from opposite sides of the property. The GMRS Base Station/Simplex Repeater sits on a corner table and is powered using AC house power (thus no need to change batteries every few days). When using GMRS handheld radios and finding the distance or terrain makes communication difficult, it is now a simple matter of entering the appropriate CTCSS tone for my simplex repeater, and communications are quickly restored.

For HAM radio I use a sturdier case, modified by adding a through-the-case coax connector. The inside of the box is then padded to protect the radio and repeater. Space is available to connect an external battery to run both the radio and simplex repeater, thereby extending operating time before it is necessary to recharge or replace batteries.

Any type of external antenna could be used with this configuration. Pictured in the accompanying photo is a folding J-pole antenna constructed out of TV ladder line and cut to operate most effectively within the 2-meter HAM radio band. Adding this external antenna greatly extends the range of the radio when compared to using a "rubber duck" whip-type antenna.

When deploying this repeater, the repeater box is placed at the base of a tree. The rope is thrown over a branch and used to pull the antenna into the tree, to the full length of the coax cable. The radio and repeater are turned on, and the radio is set to the appropriate frequency and CTCSS/PL tone. The box is then closed and left in place to operate.

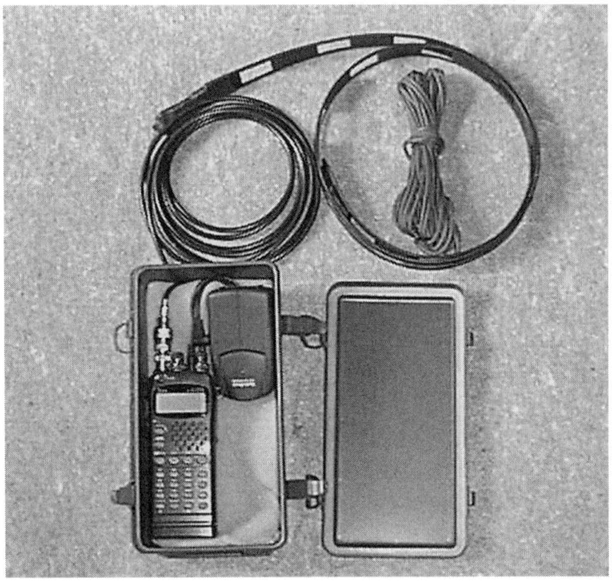

The simplex repeater prior to deployment.

Overall, simplex repeaters are excellent tools for survival and self-reliance communications. They are easy to use, they greatly increase the communications range of your personal radios, and their portability allows them to be placed wherever they are needed. As you build your personal communications package, don't forget to add a couple of simplex repeaters.

Cross-Band Repeaters

Some multiband FM transceivers have a built-in cross-band repeater function. This allows the radio to function as a duplex repeater.

If your radio has cross-band repeater

This photo shows the simplex repeater in operation. Normally the box itself would be camouflaged with natural debris to help limit accidental discovery and removal of the repeater. The antenna was not deployed to the full length of the coax, in order to allow it to be visible in the photograph.

capability, you should become familiar with this function and be able to put your radio into this mode. A mobile radio operating at 20 watts on the 2-meter band and the 440 band can serve as a cross-band repeater, greatly extending the range of handheld radios while you are away from your vehicle. If you happen to be working near a base station operating in cross-band repeater mode, the additional power and large antennas available with a base station can significantly extend the range of your handheld or mobile radios.

Wilderness Protocol

The wilderness protocol is a communications agreement among Amateur Radio operators operating radios in the UHF/VHF FM bands beyond the range of repeater systems. Originally intended to support hikers, backpackers, and others in remote areas, the wilderness protocol is very beneficial in any area lacking FM repeater support.

The wilderness protocol states that any Amateur Radio station capable of doing so should monitor the 2-meter FM calling frequency of 146.520 MHz and, if possible, the FM calling frequencies in the other UHF/VHF bands (52.525 MHz, 223.500 MHz, 446.000 MHz, and 1294.500 MHz) every three hours beginning at 0700 hours local time. Stations should listen on the primary and, if possible, the secondary frequencies for five minutes, beginning at 0700 hours, and every three hours thereafter. Stations that have sufficient power should monitor for five minutes at the top of every hour.

Stations with priority traffic should make their calls during this five-minute period, the intent being that everyone in the area with radio capability will be listening during these designated times. Calls of a general nature should not be made until the last minute of the protocol (i.e., four minutes past the hour). Of course, the wilderness protocol only becomes truly effective when the majority of radio operators in the area use it.

From a survival/self-reliance perspective, the wilderness protocol is an excellent idea. In addition to the previously mentioned Amateur Radio frequencies, I also recommend that stations having the capability to operate in the General Mobile Radio Service monitor GMRS channel 10 (the GMRS emergency channel) and stations having Citizens Band (CB) radios monitor CB channel 9 (the CB emergency channel) at the standard wilderness protocol times.

Emergency and Distress Calling

There are frequencies that are established as emergency and distress calling frequencies:

500 Khz – International Morse Code Distress **

2182 Khz – International Voice Distress and Calling

8364 Khz – International Lifeboat and Survival Craft

40.5 MHz – U.S. Army FM Voice Distress

121.5 MHz – International Aeronautical Voice Emergency

156.8 MHz – International FM Voice Distress and Calling (Marine VHF Radio Channel 16)

243.0 MHz – Combined Military Aeronautical Emergency

** Since 1999 many navies/coast guards no longer monitor the 500 Khz Morse Code Distress Frequency.

These are intended as emergency frequencies for ships and aircraft (and military) and are guarded (monitored) by organizations such as the U.S. Coast Guard as well as ships at sea and aircraft in flight. Ships and aircraft are required to guard their appropriate emergency frequency when under way and when their radios are on and not actually being used for other communications. Even if I am not acting as a radio operator aboard a ship or aircraft, I keep these frequencies programmed into a bank on my radio scanners, thus keeping a constant guard on them whenever my radios are on.

If you are in distress and need to send an emergency communication, you should do so following a fairly standard format. First and foremost, be sure that you are actually justified in sending an "in distress" communication. "In distress" is defined as any condition wherein you or your vessel is in grave and imminent danger and you require immediate assistance to preserve life or property.

Following are the U.S. Coast Guard's instructions for sending a distress call, as published on their Web site at www.navcen.uscg.gov/marcomms/boater.htm:

> Mayday! Mayday! Mayday!
>
> Sending a distress call
>
> You may only have seconds to send a distress call. Here's what you do.
>
> Transmit, in this order:
>
> If you have an MF/HF radiotelephone tuned to 2182 Khz, send the radio telephone alarm signal if one is available. If you have a VHF marine radio, tune it to channel 16. Unless you know you are outside VHF range of shore and ships, call on channel 16 first.
>
> Distress signal "MAYDAY," spoken three times.
>
> The words "THIS IS," spoken once.
>
> Name of vessel in distress (spoken three times) and call sign or boat registration number, spoken once.
>
> Repeat "MAYDAY" and name of vessel, spoken once.
>
> Give position of vessel by latitude or longitude or by bearing (true or magnetic, state which) and distance to a well-known landmark such as a navigational aid or small island, or in any terms which will assist a responding station in locating the vessel in distress. Include any information on vessel movement such as
>
> - Course, speed, and destination
> - Nature of distress (sinking, fire etc.)
> - Kind of assistance desired
> - Number of persons on board
>
> Any other information which might facilitate rescue, such as length or tonnage of vessel, number of persons needing medical attention, color hull, cabin, masks, etc.
>
> The word "OVER."
>
> Stay by the radio if possible. Even after the message has been received, the Coast Guard can find you more quickly if you can transmit a signal on which a rescue boat or aircraft can home.
>
> For example:
>
> MAYDAY-MAYDAY-MAYDAY
>
> THIS IS BLUE DUCK-BLUE DUCK-BLUE DUCK WA1234
>
> CAPE HENRY LIGHT BEARS 185 DEGREES MAGNETIC-DISTANCE 2 MILES
>
> STRUCK SUBMERGED OBJECT
>
> NEED PUMPS-MEDICAL ASSISTANCE, AND TOW
>
> THREE ADULTS, TWO CHILDREN ON BOARD
>
> ONE PERSON COMPOUND FRACTURE OF ARM
>
> ESTIMATE CAN REMAIN AFLOAT TWO HOURS
>
> BLUE DUCK IS THIRTY TWO FOOT CABIN CRUISER-WHITE HULL-BLUE DECK HOUSE
>
> OVER
>
> Repeat at interval until an answer is received.

If you hear a distress call DO NOT answer immediately. Listen carefully to the call and copy down information from it. You are listening for a response from an official agency such as the Coast Guard and determining whether you are in a position to provide assistance. It does the person in distress little good for you to answer his MAYDAY call if there is nothing you can do to help him. If you don't hear a response from the Coast Guard or other agency in a position to provide assistance to the person in distress, you are then required to answer his call. In this case, you should provide a relay of his MAYDAY call by repeating it exactly as he sent it, but in place of MAYDAY you will say MAYDAY- Relay (for example, MAYDAY-Relay – MAYDAY-Relay – MAYDAY-Relay. THIS IS BLUE DUCK-BLUE DUCK-BLUE DUCK WA1234 . . ., etc.) The assumption is that the sender of the MAYDAY has a weak or poorly operating transmitter and is not being heard by the Coast Guard or other rescue service. When the Coast Guard answers your MAYDAY-Relay, follow their instructions regarding the relay of communications.

Some people think it is funny to send false distress signals. These are the same type of people who will activate fire alarms when there is no fire. They are criminals, putting the lives of rescuers at risk and distracting rescue personnel from actual emergencies as they attempt to sort out the false distress call.

Sending a false distress message is a very serious offense:

Title 14 U.S.C. Section 88(C)

An individual who knowingly and willfully communicates a false distress message to the Coast Guard or causes the Coast Guard to attempt to save lives and property when no help is needed is -

(1) guilty of a class D felony;

(2) subject to a civil penalty of not more than $5,000; and

(3) liable for all costs the Coast Guard incurs as a result of the individual's action.

You should never hesitate to send a distress signal if you are actually in imminent danger, nor should you hesitate to relay a MAYDAY signal that isn't being heard. However, never, under any circumstances, should you send a false distress signal.

Alaska Emergency Frequency

Part 97 Section 401(d) of the FCC rules and regulations states, "A station in, or within 92.6 km of, Alaska may transmit emissions J3E and R3E on the channel at 5.1675 MHz for emergency communications. The channel must be shared with stations licensed in the Alaska-private fixed service. The transmitter power must not exceed 150 W."

Alaska is a large state with extensive remote areas. People living in these remote areas may not have landline telephone service and are often beyond the range of cellular telephone service. Because of this, the Alaska Emergency Frequency was established to allow people in remote areas to contact emergency service personnel when necessary. The frequency is monitored by police communications centers throughout Alaska, as well as other official and semiofficial organizations in the state.

The best-quality radios intended for field use, such as the outstanding Yaesu FT-817 and Yaesu FT-897, are designed to operate on the Alaska Emergency Frequency. The Alaska Emergency Frequency is not authorized beyond 92.6 km of the Alaska state borders, and would likely be of little use elsewhere, since nobody is likely to monitor the frequency outside of Alaska. However, the frequency is important if you live or travel in Alaska and should be part of your survival and self-reliance communications planning.

Maritime Mobile Service Net

The Maritime Mobile Service Net (www.mmsn.org) was established in 1968 by retired U.S. Navy Chaplain A.W. Robertson. The purpose of the Maritime Mobile Service Net is to provide legal third-party communication for maritime mobile stations and deployed military personnel. The Maritime Mobile Service Net also provides communication for missionaries in foreign lands.

The Maritime Mobile Service Net operates daily from 12 P.M. until 9 P.M. (U.S. Eastern Time) on Amateur Radio frequency 14.300 MHz (or alternate frequency 14.313 MHz). Volunteer net control stations throughout the United States and Canada maintain the Net, and several other stations serve as relays. This gives almost complete coverage of the Atlantic Ocean, Mediterranean Sea, Caribbean Sea, and eastern Pacific Ocean.

The Maritime Mobile Net invites any and all Amateur Radio operators to check into the net. This is a major advantage when attempting to make contact with another station. Although the Maritime Mobile Net focuses on and gives priority to maritime stations, land-based stations can also work through it to establish contact with each other. Of course if you have your HAM radio gear aboard a ship or boat, the Maritime Mobile Net is designed specifically for you.

I often monitor the Maritime Mobile Net and will check into the net whenever I am working in a remote area. The Maritime Mobile Net also serves as part of my "lost communications" procedure for remote area operations. In any case, where all other communications planning has, for whatever reason, failed, stations can check into the Maritime Mobile Net to reestablish contact. One of the major advantages of working through the Maritime Mobile Net is that the net control stations and the supporting relay stations can almost always assist in helping stations establish contact or in relaying brief messages from one station to another if they cannot make direct contact.

QRP—Low-Power Communications

QRP, or low-power communications, is an interesting part of Amateur Radio and a valuable addition to communications for survival and self-reliance. The ability to communicate using less than 5 watts output power is very useful if you are relying solely on batteries or another limited power source to run your radios.

Skill in QRP communication requires using proper antennas and effective modes of operation for low-power communication (such as Morse code) and carefully selecting times and frequencies to take advantage of the best propagation to the areas with which you want to communicate.

When compared with a 1,500-watt station, your 5-watt QRP station will have a very weak signal. Your weaker signal can be completely buried under these more powerful signals. Therefore, a station intending to receive your QRP signal will need to listen for it carefully. This is where communications planning is of great value.

On the other hand, QRP communication should not be considered something where you have to fight for every contact. I regularly communicate from the Pacific Northwest of the United States into Europe, Asia, and Africa using a 5-watt radio and a wire antenna. Many of these contacts are voice communication, which is much less efficient than Morse code or the various other digital modes.

QRP operators have suggested a Morse code

QRP Calling Frequencies

Band	CW	SSB	FM
160 Meters	1.810	1.910	
80 Meters	3.560	3.985	
40 Meters	7.040	7.285	
30 Meters	10.106		
20 Meters	14.060	14.285	
17 Meters	18.096		
15 Meters	21.060	21.385	
12 Meters		24.906	
10 Meters	28.060	28.885	
6 Meters	50.060	50.885	
2 Meters	144.060	144.285	144.585

and voice calling frequency on each of the Amateur Radio bands. On and around these frequencies it is easy to find radio operators running their low-power rigs.

For the radio operator interested in survival and self-reliance communications, I recommend that he spend a good deal of time running low-power communication. You should be aware of just what your radios will do when running less than 5 watts and just how long they will do it, running off their internal batteries (or whatever external battery pack you have set up for them).

GENERAL MOBILE RADIO SERVICE

With regard to the GMRS, the FCC says, "The General Mobile Radio Service ("GMRS") is a personal radio service available to an individual (one man or one woman). It is a two-way voice communication service to facilitate the activities of the individual's immediate family members. Expect a communications range of five to twenty-five miles. You cannot make a telephone call with a GMRS unit."

GMRS has 15 channels operating in the 462 MHz range and 7 repeater input frequencies in the 467 MHz range. Maximum allowable power for GMRS radios is 50 watts; however, the small handheld units usually run at 2 to 5 watts. Communication range with GMRS radios is about 5 miles with the handheld units and perhaps 15 to 25 miles with the more powerful base/mobile units. A station license is required to operate GMRS.

GMRS/FRS combination radios are made by several of the major radio manufacturers. While most GMRS radios will contain the seven shared GMRS/FRS frequencies; it is also possible to purchase radios that also contain the other seven

GMRS Frequencies

Channel #	Frequency	
1	462.5625 MHz	
2	462.5875 MHz	
3	462.6125 MHz	
4	462.6375 MHz	
5	462.6625 MHz	
6	462.6875 MHz	
7	462.7125 MHz	
8	462.5750 MHz	
9	462.6250 MHz	
10	462.6750 MHz	Emergency Channel
11	462.5500 MHz	
12	462.6000 MHz	
13	462.6500 MHz	
14	462.7000 MHz	
15	462.7250 MHz	

GMRS Repeater Input Frequencies

467.5500 MHz
467.5750 MHz
467.6000 MHz
467.6250 MHz
467.6750 MHz
467.7000 MHz
467.7250 MHz

FRS only frequencies (e.g., Motorola T-7200). If you have your GMRS license (remember, this is a licensed service), you can operate on the shared GMRS/FRS frequencies at greater power output than a nonlicensed FRS operator can. The inclusion of the additional seven FRS-only frequencies in your radio allows you to operate on all FRS channels, but on the FRS-only frequencies your power output is automatically reduced 500 mW (one-half watt) in most of these radios.

GMRS has several advantages over FRS. The first and perhaps most important advantage is that you are allowed to use external antennas with your GMRS radios. The ability to mount an antenna on a rooftop or other high location provides a significant increase in transmission range of GMRS when compared to FRS. Another advantage, of course, is the greater increased output power allowed for GMRS.

Many GMRS operators also set up repeater systems allowing even greater range and communications capability. It is even possible to establish a simplex repeater (described previously in this text) for use with your GMRS radios.

For local communications, GMRS has several significant advantages. I have found that GMRS is much more effective than citizen's band (CB) radios for local personal communications. For a family or small group communications network, GMRS is an excellent choice. A GMRS base station established at home, GMRS radios mounted in your vehicles, and handheld GMRS radios all make for a well-rounded communication package.

LAND MOBILE RADIO SERVICE

The Land Mobile Radio Service is intended for use by business/industrial, transportation (railroad, taxicab, mobile carrier), and public safety organizations.

It is possible for persons engaged in certain activities to receive a channel allocation within the Land Mobile Radio Service. FCC regulations state:

> Persons primarily engaged in any of the following activities are eligible to hold authorizations in the Industrial/Business Pool to provide commercial mobile radio service . . . or to operate stations for transmission of communications necessary to such activities of the licensee:
> The operation of a commercial activity;
> The operation of educational, philanthropic, or ecclesiastical institutions;
> Clergy activities; or
> The operation of hospitals, clinics, or medical associations.

Assignment of a frequency within the Land Mobile Radio Service requires going through a frequency coordinator. If you are involved in activities eligible for a Land Mobile Radio Service frequency, you may want to apply for said frequency. However, I do not believe that there is any particular advantage to use of the Land Mobile Radio Service in communications for survival and self-reliance.

These frequencies are all UHF/VHF allocations, and are thus reasonably short range. Furthermore, you will be limited to your assigned frequency. Other radio services discussed in this book are much more applicable for our communications purposes.

Although we probably won't want to go through the effort of receiving a frequency allocation in the Land Mobile Radio Service, we will certainly want to monitor these frequencies on our scanners. Remember, public safety organizations, such as police, fire, and rescue services, have their frequencies in the Land Mobile Radio Service, as do transportation services and various public utility services.

MARITIME (MARINE) RADIO SERVICE

The Maritime (Marine) Radio Service is, as its name would imply, intended for boaters and ships on the waterways and oceans. Each channel within this radio service is intended for specific purposes, as indicated in the following chart.

To operate a marine radio, you may or may not be required to have an FCC license, depending on your specific situation. On

Channel Number	Ship Transmit MHz	Ship Receive MHz	Use
01A	156.050	156.050	Port operations and commercial, VTS. Available only in New Orleans/Lower Mississippi area.
05A	156.250	156.250	Port Operations or VTS in the Houston, New Orleans, and Seattle areas.
06	156.300	156.300	Intership safety.
07A	156.350	156.350	Commercial.
08	156.400	156.400	Commercial (intership only).
09	156.450	156.450	Boater calling. Commercial and noncommercial.
10	156.500	156.500	Commercial.
11	156.550	156.550	Commercial. VTS in selected areas.
12	156.600	156.600	Port operations. VTS in selected areas.
13	156.650	156.650	Intership navigation safety (bridge-to-bridge). Ships >20m length maintain a listening watch on this channel in U.S. waters.
14	156.700	156.700	Port operations. VTS in selected areas.
15	—	156.750	Environmental (receive only). Used by Class C EPIRBs.
16	156.800	156.800	International distress, safety, and calling. Ships required to carry radio, USCG, and most coast stations maintain a listening watch on this channel.
17	156.850	156.850	State control.
18A	156.900	156.900	Commercial.
19A	156.950	156.950	Commercial.
20	157.000	161.600	Port operations (duplex).
20A	157.000	157.000	Port operations.
21A	157.050	157.050	U.S. Coast Guard only.
22A	157.100	157.100	Coast Guard liaison and maritime safety. Information broadcasts. Broadcasts announced on channel 16.
23A	157.150	157.150	U.S. Coast Guard only.
24	157.200	161.800	Public correspondence (marine operator).
25	157.250	161.850	Public correspondence (marine operator).
26	157.300	161.900	Public correspondence (marine operator).
27	157.350	161.950	Public correspondence (marine operator).
28	157.400	162.000	Public correspondence (marine operator).

63A	156.175	156.175	Port operations and commercial, VTS. Available only in New Orleans/Lower Mississippi area.
65A	156.275	156.275	Port operations.
66A	156.325	156.325	Port operations.
67	156.375	156.375	Commercial. Used for bridge-to-bridge communications in lower Mississippi River. (intership only).
68	156.425	156.425	Noncommercial.
69	156.475	156.475	Noncommercial.
70	156.525	156.525	Digital selective calling (voice communications not allowed).
71	156.575	156.575	Noncommercial.
72	156.625	156.625	Noncommercial (intership only).
73	156.675	156.675	Port operations.
74	156.725	156.725	Port operations.
77	156.875	156.875	Port operations (intership only).
78A	156.925	156.925	Noncommercial.
79A	156.975	156.975	Commercial. Noncommercial in Great Lakes only.
80A	157.025	157.025	Commercial. Noncommercial in Great Lakes only.
81A	157.075	157.075	U.S. government only–environmental protection operations.
82A	157.125	157.125	U.S. government only.
83A	157.175	157.175	U.S. Coast Guard only.
84	157.225	161.825	Public correspondence (marine operator).
85	157.275	161.875	Public correspondence (marine operator).
86	157.325	161.925	Public correspondence (marine operator).
87	157.375	161.975	Public correspondence (marine operator)
88	157.425	162.025	Public correspondence only near Canadian border.
88A	157.425	157.425	Commercial, intership only.

October 26, 1996, the FCC eliminated the license requirement for boats voluntarily carrying VHF marine radios and operating only within U.S. waters (i.e., no trips to Canada, Mexico, or other foreign ports). Basically this means that for your personal pleasure craft you probably won't need an FCC radio license to operate your VHF marine radio, EPIRB, and ship radar.

You will require an FCC radio license if your vessel is required by law to carry radio equipment, if you travel to foreign ports, if your craft is greater than 20 meters in length, if you carry more than six passengers for hire, and under various other conditions, or if you intend to use HF/MF radios or satellite communications.

Whether you are required to have an FCC radio license or not, if you operate a boat or ship larger than a rowboat, VHF marine radios are an essential part of your survival equipment.

MILITARY AFFILIATE RADIO SYSTEM

The official definition of MARS is, "MARS is a Department of Defense sponsored program, established as a separately managed and operated program by the Army, Navy, and Air Force. The program consists of licensed Amateur Radio operators who are interested in military communications on a local, national, and international basis as an adjunct to normal military communications." The mission of MARS is summed up in three basic statements:

> To provide Department of Defense (DOD)-sponsored emergency communications on a local, national, and international basis as an adjunct to normal communications.
>
> To provide auxiliary communications for military, civil, and/or disaster officials during periods of emergency.
>
> To assist in effecting normal communications under emergency conditions.

Any Amateur Radio operator can apply to work with MARS. MARS works on military frequencies, so it will be necessary to make modifications to most HAM radio equipment, allowing it to transmit outside of the Amateur bands. Fortunately these modifications are not particularly difficult, and the major radio manufacturers of Amateur Radio equipment will provide instructions for modifying your radios once you are part of the MARS system.

If you are interested in participating in MARS, you may contact MARS for each military service at the following addresses:

Air Force:
HQ AFCA/GCGS (MARS)
203 W Losey Street, Room 3065
Scott AFB, IL 62225-5222

Army:
Attn: NETC OPE MA
U.S. Army NETCOM/9th ASC
Suite 3102
Fort Huachuca, AZ 85613-7070

Navy/Marine Corps:
Nebraska Avenue Complex
4234 Seminary Drive NW, Suite 19239
Washington, DC 20394-5461

CHAPTER 3

Unlicensed Radio Services

Now let's take a look at the various unlicensed personal radio services in the United States. By unlicensed personal radio services, we are talking about those that allow you to simply purchase a radio, turn it on, and begin transmitting without any requirement for any type of station or operator licensing.

CITIZENS BAND RADIO

In the United States, CB radio operates on 40 AM channels, between 26.965 MHz and 27.405 MHz, with a maximum power output of 4 watts. In addition to the 40 AM channels, operation is authorized on the upper and lower side band of these channels with a maximum power output of 12 watts.

CB radio has been in existence since the 1950s, and since that time there have been literally millions of CB radios sold to the general public. Most any trucker will have a CB installed in his truck, and in some areas it seems like every other vehicle has a CB radio installed. Beyond mobile rigs, lots of people enjoy CB radio as a hobby and have base stations and large antenna systems set up in their homes. There are also handheld (walkie-talkie) CB radios used by hunters, backpackers, and others needing portable short-range communication. Simply put, you will be hard-pressed to find any populated area within the United States where CB operators are not active on the airwaves.

From a survival communications point of view, this can be both good and bad. Clearly, having a large segment of the population using a communications system that is not dependent on the local infrastructure can be invaluable for gathering information or summoning assistance during an emergency. CB radios are also a useful aid while traveling by automobile to stay abreast of local traffic road conditions or situations that could affect the flow of traffic in a particular area.

One of the disadvantages of CB radio is that in some areas the channels tend to be crowded. With only 40 authorized channels and the large number or people owning and operating CB radios, it can be difficult to find an open channel at times. Overall, this isn't an overwhelming problem, but it is something to be aware of, especially if you intend to use your CB radio as a primary means of communication during a disaster or widespread emergency situation.

Unfortunately, with the popularity of CB radios come those individuals who believe that

CB radio is their ticket to act like complete idiots on the airwaves. These are the same individuals who transmit noise or music with the intent of jamming a channel, operate with excessive power, and intentionally cause harmful interference to others trying to use CB radios to communicate. All of these actions are illegal, but CB radio is, for all practical purposes, unregulated. The Federal Communications Commission (FCC) has clearly established regulations for the use of CB radio but doesn't bother to enforce them, and in contrast to Amateur Radio, there is little to no self-regulation and self-policing among CB radio operators.

CB Frequencies (Channel - MHz):

1- 26.965	11- 27.085	21- 27.215	31- 27.315
2- 26.975	12- 27.105	22- 27.225	32- 27.325
3- 26.985	13- 27.115	23- 27.235	33- 27.335
4- 27.005	14- 27.125	24- 27.245	34- 27.345
5- 27.015	15- 27.135	25- 27.255	35- 27.355
6- 27.025	16- 27.155	26- 27.265	36- 27.365
7- 27.035	17- 27.165	27- 27.275	37- 27.375
8- 27.055	18- 27.175	28- 27.285	38- 27.386
9- 27.065 *	19- 27.185	29- 27.295	39- 27.395
10- 27.075	20- 27.205	30- 27.305	40- 27.405

* Emergency & Travelers' Assistance

Those causing harmful interference and disrupting normal operations on CB radio tend to give CB radio operators a bad name among operators in other radio services (especially Amateur Radio). However, it is important to recognize that these criminal operators are not in the majority and that most people using CB radios are conscientious, well-mannered individuals who are using their radios for legitimate communications.

Unfortunately, illegal operations on CB frequencies and the complete lack of FCC enforcement or self-regulation can make CB radio of limited value as a means of reliable survival and self-reliance communication in many populated areas. The greater the number of people in any given area trying to use a limited 40 channels, the greater the probability of some idiot's intentionally disrupting communications.

However, I still recommend that you add a CB radio to your communications planning. The fact that a large segment of the population has and uses (or at least has access to) CB radios makes them a very useful tool for local communication.

There are a few things to consider when purchasing a CB radio as part of your overall communications package. First, buy a top-quality radio made buy a well-established manufacturer. There are certainly inexpensive CB radios on the market, but poor-quality equipment gives poor reception and transmission capability. Furthermore, the better-quality radios usually contain noise-filtering circuitry that might not be present in the less expensive lower-quality equipment.

Secondly, buy a CB with side band (SSB) capability. There is less congestion and deliberate interference on the SSB frequencies (fewer CB operators have SSB radios), and the increased power (12W vs. 4W) allowed on SSB helps get your signals through. There are two side bands to every AM channel on your CB: USB (upper side band) and LSB (lower side band).

Finally, get the best antenna available with the resources you have available. Even the best radio will perform poorly with a poor antenna.

I generally recommend directional, "beam-type" antennas for CB base stations. This allows you to direct your signal toward your intended receiving station while at the same time limiting interference from other signals to the sides and rear of the intended signal path. I have a quality CB base station combined with a directional antenna and maintain regular contact with another similarly set-up CB base station at a measured distance of 117 miles. We simply determined the azimuth between the two stations and pointed our directional antennas along this azimuth. We operate on a given SSB frequency at the legal 12W output and have clear, strong signals between the two stations.

Base-to-base communication is a good use of CB for survival and self-reliance communications. CB radio operates on the 11-meter band (between the 10- and 12-meter Amateur Radio bands). Propagation on 11 meters is best during daylight hours, but the frequencies are certainly useable at night for local communications. The FCC restricts CB communication to about 150 miles, which is about all one can legitimately expect to do on a regular basis using legal equipment and staying within assigned power restrictions. The potential for worldwide communication with CB radio is within the propagation potential of the band but cannot be relied on and thus should not be part of your communications plan.

Although CB radio is very popular as a mobile (in a vehicle) radio, for survival and self-reliance communications, they are of somewhat limited value. Mobile radios don't have the advantage of using large or directional antennas to direct signals toward the intended receiving station and limit outside interference. Some have recommended using a mobile CB simply to monitor signals in a given area or as a way to learn of road conditions when traveling. There is some merit to this suggestion, but I have found that CB frequencies can be programmed into a scanner if one wants to just monitor for possible information. Use of a scanner also allows local police, fire, and emergency service frequencies to be programmed, which will usually provide more information than CB radio frequencies.

One exception I make to using mobile CB radio is when there is an active REACT group in your area and/or when you have an established base station that will listen for your signals and respond to them. When Radio Emergency Associated Communication Teams (REACT) or a similar group is active in an area, monitoring CB channel 9, and able to give assistance to those calling on that channel, it may be worth having mobile CB capability, especially if you don't have any other radio capability.

REACT

REACT was formed in 1962 with the idea of using CB radio to provide a means of organized emergency radio communication for the general public. The REACT mission statement says, "We will provide public safety communications to individuals, organizations, and government agencies to save lives, prevent injuries, and give assistance wherever and whenever needed." Truly a worthwhile goal.

REACT was instrumental in getting the FCC to designate CB channel 9 as an emergency channel in the early 1970s. Where REACT groups are active, they monitor CB channel 9 in order to provide emergency assistance and traveler aid to those in need. This can be a useful resource for survival communications. Unfortunately, REACT is not active in all areas.

If REACT is not active in your particular area, and you have an interest in forming a group, REACT encourages you to do so. REACT may be contacted at

REACT International, Inc.
5210 Auth Rd. #403
Suitland, MD 20746

E-mail: react@reactintl.org
Web site: www.reactintl.org

It is interesting to note that while we often tend to think of REACT as CB radio-associated organization, they do not limit themselves to CB. Rather, they use all means of communication (Amateur Radio, General Mobile Radio Service, Cellular Telephone, and so on) for which a particular REACT group has the appropriate equipment and licenses.

Freeband Radio

Freeband radio is the illegal use of frequencies within the 11-meter band that are not assigned to the citizen's band radio service. Because freeband radio is not assigned operating frequencies, there is some debate as to exactly what the freeband frequencies are, but one may generally say that freeband radio ranges from 26.000 MHz up to 26.965 MHz, which is channel 1 in the assigned citizen band frequencies. Freeband then goes from 27.405 MHz (CB channel 40) up to 28.000 MHz, the beginning of the Amateur Radio 10-meter band.

Freeband radio enjoys worldwide popularity. There are even freeband radio clubs accepting membership and assigning these members freeband radio call signs. These clubs run their own radio nets and often work at making long-range/international radio contacts—referred to by radio operators as "DX," or "shooting the skip."

The major problem with freeband radio is that *it is illegal in the United States*, and in many other countries as well! Although there are areas of the 11-meter band that are not assigned to the citizen's band radio service, this does not mean that these frequencies are completely unallocated. Freeband operators may be interfering with agencies assigned frequencies within the 11-meter band outside of the CB frequencies. Moreover, freeband operators tend to stray into the Amateur Radio 10-meter band above 28.000 MHz, thereby causing harmful interference to the Amateur Radio Service.

UK Citizen's Band Radio

In the United Kingdom, CB radios are allocated 80 channels. The first 40 channels are the same as in the United States (26.965 – 27.405 MHz). The next 40 channels (channels 41 through 80) are allocated in the frequency range 27.6 – 27.99 MHz.

CB radios designed for use in the UK operate in the FM mode. CB radios designed to operate in the United States operate in the AM mode. This unfortunately makes a "UK CB" and a "U.S. CB" incompatible. The FM and AM modes cannot talk to each other, and trying to listen to an FM signal with an AM receiver (and vice versa) results in a very poor-quality signal.

Due to the FM mode of operation and additional channels of UK CBs, some individuals have ordered them for use in the United States. One major supplier of CB radios in the UK that will export to the United States is Knights CB:

Knights CB
11 King Edward St.
Kirton in Lindsey
Lincolnshire. DN21 4NF UK
E-mail: info@kcb.co.uk
Web site: www.kcb.co.uk

These individuals using CB radios designed for use in the UK in the United States are seeking expanded operating capability and perhaps the ability to escape the overcrowding and poor operating procedures found on legal CB in the United States. However, like freeband radio, the use of CB radios intended for use in the UK is illegal in the United States.

FAMILY RADIO SERVICE (FRS)

The Family Radio Service, or FRS radio, is a UHF/FM radio service intended for communication within families and other like groups. It is specifically NOT intended for business or commercial use. FRS is limited, by FCC regulations, to a maximum output power of 500mW (one-half watt).

FRS is authorized 14 channels within the General Mobile Radio Service (GMRS) frequency range. In fact the first seven FRS and GMRS channels are the same.

FRS Frequencies

1	462.5625 MHz	8	467.5625 MHz
2	462.5875 MHz	9	467.5875 MHz
3	462.6125 MHz	10	467.6125 MHz
4	462.6375 MHz	11	467.6375 MHz
5	462.6625 MHz	12	467.6625 MHz
6	462.6875 MHz	13	467.6875 MHz
7	462.7125 MHz	14	467.7125 MHz

With the 500mW power limitation and a prohibition against using external antennas, FRS is limited to an effective communication range of less than two miles, with about one mile being the average. However, this short-range communication is not necessarily a bad thing. For communication among a small group working in the same general area, FRS is ideal. The fact that FRS has limited range means that you are not broadcasting your communications across half the state when all you need to do is talk with someone a mile or so away.

Many FRS radios are equipped with something billed as "38 privacy codes" or the like. These privacy codes do not, as some believe, prevent others from hearing your conversation unless they have the same code set on their radio. These privacy codes are CTCSS (continuous tone coded squelch system), which keeps you from hearing signals unless the signal includes a matching CTCSS tone. Someone with no privacy code set will hear all traffic on a given channel. If you have a privacy code set, you will only hear transmissions including that particular code, the idea being that you will only hear transmissions intended for you.

If you purchase radios from different manufacturers, you should be aware that the CTCSS tones used by one manufacturer might not be compatible with those used by a different manufacturer. The following chart lists all of the CTCSS tones as recognized by the Electronic Industries Association, listing their hertz tones and their Motorola alphanumeric designations.

Continuous Tone Coded Squelch System

67.0 - XZ	110.9 - 2Z	186.2 - 7Z
69.3 - WZ	114.8 - 2A	192.8 - 7A
71.9 - XA	118.8 - 2B	203.5 - M1
74.4 - WA	123.0 - 3Z	203.5 - M1
77.0 - XB	127.3 - 3A	206.5 - 8Z
79.7 - WB	131.8 - 3B	206.5 - 8Z
82.5 - YZ	136.5 - 4Z	210.7 - M2
85.4 - YA	141.3 - 4A	218.1 - M3
88.5 - YB	146.2 - 4B	225.7 - M4
91.5 - ZZ	151.4 - 5Z	229.1 - 9Z
94.8 - ZA	156.7 - 5A	233.6 - M5
97.4 - ZB	162.2 - 5B	241.8 - M6
100.0 - 1Z	167.9 - 6Z	250.3 - M7
103.5 - 1A	173.8 - 6A	254.1 - 0Z
107.2 - 1B	179.9 - 6B	

If you take the time to count the CTCSS tones in the above chart, You will see that there are more than 38 tones. These CTCSS tones are used in all manner of radios. On my HAM radio equipment I can program any CTCSS tone by the actual hertz used. On an FRS radio, the manufacturer typically chooses a number of these tones and allows you to select them by number, but not by the actual hertz itself. This can lead to different hertz tones being assigned to the same number by different manufacturers. For example, on my Cherokee FRS radio, "privacy code" number 2 is 69.3 hertz (Hz), whereas it is 71.9 Hz on my Motorola FRS radio and 69.4 Hz on my Radio Shack FRS radio. This is not an overwhelming problem, since the owner's manual for each of the radios lists the actual hertz tone assigned to the "privacy code" numerical designation, but it is something to be aware of when using radios from different manufacturers.

FRS is very useful for family outings. Anytime a small group is conducting an activity where someone may become separated from the group, whether intentionally or accidentally, FRS keeps everyone in contact. No matter what type of radios you choose as the basis for your survival and self-reliance communications plan, I recommend that it include a quality FRS radio for every member of your group.

FRS Neighborhood Watch Net

A very useful program using FRS radios is the FRS Neighborhood Watch Net. Setting up an FRS Neighborhood Watch Net is as simple as making an agreement among families in a neighborhood to regularly monitor a selected FRS channel and to answer calls for assistance on that channel.

While any type of FRS radio may be used to participate in the FRS Neighborhood Watch Net, I recommend one of the various FRS "base stations," simply because these radios are powered by plugging them in to a common AC outlet, thus precluding the need for a large supply of batteries. Of course, during a power outage an FRS walkie-talkie would work just as well, and people outside of their homes may choose to carry an FRS walkie-talkie to have access to the FRS Neighborhood Watch Net.

Once the FRS Neighborhood Watch Net is established, it should be tested once per week. This need not be a very complicated procedure. The net controller for the week simply calls the net, asks for check-ins, asks for traffic, and closes the net. For example:

Net controller: "Good Evening. This is the weekly Wednesday-night test of the FRS Neighborhood Watch Net. Anyone wishing to check in to the net, please come now with your name."

"Smith family checking in."
"John Jones."
"Wilson family."

Net controller: "We have three check-ins tonight. Does anyone have anything for the net this evening?"

(After anyone with traffic for the net has finished, the net controller closes the net.)

Net controller: "Thank you all for checking in tonight. John Jones has agreed to act as net controller next week. See you all on the net next Wednesday at 7 P.M. This completes the FRS Neighborhood Watch Net test for this week."

The FRS Neighborhood Watch Net need not be overly formal, nor is it just for emergencies. It keeps local neighbors in touch, providing quick communication in a neighborhood in case there is some kind of problem or someone needs assistance. If you already have a neighborhood watch program established in your neighborhood, adding FRS radios can only improve it.

Because FRS channels are shared by all users, it might be of value to assign a CTCSS privacy code to the net's channel. This helps prevent the occasional transmission not intended for the Neighborhood Watch on the frequency used by the net from being received by the entire neighborhood.

The FRS Neighborhood Watch Net is an excellent example of communications for self-reliance: neighbors helping neighbors.

MULTI-USE RADIO SERVICE (MURS)

On November 13, 2000, the FCC designated the Multi-Use Radio Service (MURS) as a nonlicensed allocation of frequencies. MURS uses five VHF frequencies from the old "business radio" allocation. Use of the MURS is limited to two watts output power but has no limitations on external antennas and such, as does FRS.

The MURS frequencies are

- 151.8200 MHz
- 151.8800 MHz
- 151.9400 MHz
- 154.5700 MHz
- 154.6000 MHz

The major radio manufacturers are beginning to produce radios that operate on the five MURS radio frequencies. One example of this is the Maxon MURS25 radio. Radios such as the Motorola Spirit and Radius, with their Blue Dot Frequency (154.5700 MHz) and Green Dot Frequency (154.6000 MHz), operate in what is now MURS. Radio Shack offered a two-channel VHF Radio, BTX-127, for a while that operated on these frequencies. I note that the BTX-127 is no longer available on the Radio Shack Web site but can still be found in stock in some Radio Shack stores. Radio Shack does, however, currently offer a two-channel mobile radio that can be programmed with MURS frequencies.

The MURS radios operating in the VHF range provide better communication in areas where signals must bend over hills and penetrate into areas with trees and vegetation. The additional power authorized to MURS radios when compared to FRS radios (2 watts on MURS compared with one-half watt on FRS) also adds a bit to the communications range. However, you will find that your range with MURS radios in a portable mode (walkie-talkie) is limited to about 3 to 5 miles, and perhaps a bit more than that using external antennas from a mobile or base station MURS radio.

MURS is an improvement over FRS for mobile and base stations, specifically because of the ability to use external antennas. However, because MURS is limited to only five channels, the potential for overcrowding is fairly significant. In rural and remote areas, where there is little or no other MURS activity, the MURS system provides an advantage over FRS. To determine if MURS is heavily used in your area, simply program the five MURS frequencies into your scanner and listen for several days. If you hear little or no traffic, a MURS radio system might be something to consider for establishing a communications network for you and your friends in that area. After all, if nobody else is using the frequencies, you might as well take advantage of the free bandwidth.

49 MHZ RADIOS

The 49 MHz radios are intended for very short-range, two-way communication. These radios are most often seen as a belt unit and headset/microphone combination, although some children's toy walkie-talkie radios operate on 49 MHz. The range of these radios is somewhere between 100 yards and a quarter mile, depending on the quality of the radio and the terrain in which they are being used.

These radios, in their hands-free headset configurations, are useful for communications while traveling by motorcycle, climbing, and hunting. From the survival viewpoint, the advantage of 49 MHz radios is that they allow communication between people working in close proximity to each other without broadcasting that communication to others outside of the group.

The 49 MHz radio frequencies are

- 49.830 MHz
- 49.845 MHz
- 49.860 MHz
- 49.875 MHz
- 49.890 MHz

Of course, the disadvantage of the 49 MHz radios is their very short communications range. Basically, if you are much beyond a quarter mile from the person with whom you are trying to communicate, you will have problems using commonly available models.

CHOOSING AN UNLICENSED RADIO SERVICE

After looking at the various unlicensed personal radio services available, the question invariably arises as to which one is best. As with beauty, best is in the eye of the beholder, but we can make some general comparisons to help you choose the radio that will best meet your needs.

First, let us consider the 49 MHz radios. These are best used for only very short-range communications (a couple hundred yards or so). In their hands-free/headset designs, they come in handy for communication while biking, hiking, and other like activities. They are also handy for talking from a garage/workshop into the associated home or for communication between closely situated homes or apartments. Generally, I do not see any particular advantage to 49 MHz radios for survival and self-reliance communications.

Next we will consider CB, FRS, and MURS. The FCC describes each of these personal radio services as follows:

> CB is one of the Citizens Band Radio Services. It is a two-way voice communication service for use in your personal and business activities. Expect a communication range of one to five miles.

> The Family Radio Service (FRS) is one of the Citizens Band Radio Services. It is for your family, friends, and associates to communicate among yourselves within your neighborhood and while on group outings. You cannot make a telephone call with a FRS unit. You may use your FRS unit for business-related communications. Expect a communication range of less than one mile.

> The Multi-Use Radio Service (MURS) is an unlicensed, five-channel personal radio service implemented in fall 2000 in the VHF 150 MHz band. It uses low-power handheld radios on frequencies formerly available only for licensed business and commercial operations. (Personal Radio Steering Group [PRSG] definition)

As these descriptions clearly convey, the unlicensed radio services are intended for short-range personal communication. When choosing one of these radio services, it is important to be aware of their limitations. For example, you should not expect to communicate over a distance of 25 miles on your CB radio, since the expected effective communications range is only one to five miles. It is certainly possible to communicate 25 miles or more with a CB radio, but such ranges should not be part of your communication plan, unless you have demonstrated that ability with your particular equipment repeatedly.

FRS radios all advertise communication ranges up to two miles, "up to" being the operative words here. Under perfect communications conditions you might be able to have reliable communications with someone two miles distant, but don't count on it under normal circumstances.

MURS radios are a fairly new unlicensed service (having come from the old business radio service). Assuming you don't have a local business still trying to use a MURS frequency for business purposes, MURS makes an excellent addition to your communications package. This is especially true for use as mobile radios. You shouldn't expect communications ranges greater than about five miles, but the MURS frequencies operating in the VHF range are quiet and usually quite clear when compared to CB.

If using only the unlicensed radio services, I recommend CB radios for general communication (it seems that everyone has a CB), MURS for mobile and small base station setups, and FRS for handheld person-to-person communication on outings or when working in a limited area. Using the unlicensed radio services, one can have reasonable communications with friends and family in a limited area, but to have the most effective communications it will be necessary to move up to the licensed radio services.

CHAPTER 4

Operating Procedures

Operating procedures provide a standard format for sending and receiving radio communications. They help ensure accuracy and reduce ambiguity in communication.

COORDINATED UNIVERSAL TIME (UTC)

Because our communications may extend across several states or even around the world, it is important to be able to have a means of standardizing time. If you live in Texas and tell your friends in California and New York that you should all meet on the air at 10:00 on Tuesday, you might find that none of you are on the air at the same time.

There is an hour difference between 10:00 in Texas and California, two hours difference between Texas and New York, and three hours difference between California and New York. Add a friend in Australia into this conversation and you might not even have everyone on the air on the same day!

Because of these different time zones, we use universal time in all communications planning. Universal time is also known as Coordinated Universal Time (usually abbreviated UTC from the French Universelle Tempes Coordinate), Greenwich Mean Time (GMT), or Zulu Time (Z) from the time zone at longitude zero (the prime meridian), which passes through Greenwich, England.

To coordinate accurate communication schedules, it is necessary that they be established based on a standardized time. Generally, we will always refer to Coordinated Universal Time when discussing communications, but we still have the problem of ensuring that everyone sets their clocks to the correct UTC time. Looking at your wristwatch and then adding or subtracting the appropriate number of hours works only if your wristwatch is set accurately.

In order to have a time standard for all communications planning, we can use the National Institute of Standards and Technology (NIST) official clock. Official time is available on the Internet at www.time.gov.

Additionally, NIST broadcasts time announcements 24 hours per day, from station WWV in Fort Collins, Colorado, on 2.5 MHz, 5 MHz, 10 MHz, 15 MHz, and 20 MHz. NIST radio station WWVH on Kauai, Hawaii, broadcasts these same time announcements, but only on 2.5 MHz, 5 MHz, 10 MHz, and 15 MHz.

Using the NIST Internet clock, or the NIST radio time broadcasts will allow accurate synchronization of time standards among all stations participating in a communications network.

PHONETIC ALPHABET

The phonetic alphabet is an internationally accepted means of ensuring accuracy in spelling during voice communication via radio and telephone. On a noisy circuit it may be difficult to tell whether someone said "A" or "I," but there is much less difficulty in distinguishing between the words "Alpha" and "India." You might mistake the pronunciation of "B" for "P," but it is unlikely that you would confuse the word "Bravo" for "Papa." All competent radio operators learn the phonetic alphabet and use it to avoid mistakes in spelling during voice communication.

A – Alpha	N – November
B – Bravo	O – Oscar
C – Charlie	P – Papa
D – Delta	Q – Quebec
E – Echo	R – Romeo
F – Foxtrot	S – Sierra
G – Golf	T – Tango
H – Hotel	U – Uniform
I – India	V – Victor
J – Juliet	W – Whiskey
K – Kilo	X – X-Ray
L – Lima	Y – Yankee
M – Mike	Z – Zulu

Q-CODES

The Q-Codes, or Q-Signals, are three-letter groups, with each said group beginning with the letter Q. The Q-Codes were established at the Radiotelegraph Convention of 1912 in London, England. The Q-Codes are divided into four basic groups: QAA – QNZ for aeronautical service, QOA – QQZ for maritime service, QRA – QUZ for general service, and QZA – QZZ for special services and other uses.

We will concern ourselves with the QRA – QUZ Q-Codes used in general service, or, more specifically, those used by radio operators. In actual practice you will likely use no more than 8 or 10 of the Q-Codes on a regular basis, although I have included others that you might hear and use from time to time.

The Q-Codes are primarily intended for use in Morse code communication, although you will hear some of them used in voice communication (e.g., "My QTH is Boulder, Colorado.")

The main advantage of using the Q-Codes is that they allow for abbreviated signals and communication of basic information between people who do not speak the same language.

QRA What is the name of your station?
The name of my station is ___ .

QRB What is the approximate distance between our stations?
The approximate distance between our stations is ___.

QRD Where are your bound for, and where are you from?
I am from ___ and bound for ___.

QRG Will you tell me my exact frequency (or that of ___)?
Your exact frequency (or that of ___) is ___ Khz.

QRH Does my frequency vary?
Your frequency varies.

QRI How is the tone of my transmission?
The tone of your transmission is ___ (1. Good; 2. Variable; 3. Bad).

QRJ Are you receiving me badly?
I cannot receive you. Your signals are too weak.

QRK What is the intelligibility of my signals (or those of ___)?
The intelligibility of your signals is ___ (1. Bad; 2. Poor; 3. Fair; 4. Good; 5. Excellent).

QRL Are you busy?
I am busy (or I am busy with ___).
Please do not interfere.

QRM Is my transmission being interfered with?
Your transmission is being interfered with ___ (1. Nil; 2. Slightly; 3. Moderately; 4. Severely; 5. Extremely).

QRN Are you troubled by static?
I am troubled by static ___. (Usually, QRM is man-made interference and QRN is natural interference.)

QRO Shall I increase power?
Increase power.

QRP Shall I decrease power?
Decrease power.

QRQ Shall I send faster?
Send faster (___ WPM).

QRS Shall I send more slowly?
Send more slowly (___ WPM).

QRT Shall I stop sending?
Stop sending.

QRU Have you anything for me?
I have nothing for you.

QRV Are you ready?
I am ready.

QRX When will you call me again?
I will call you again at ___ hours (on ___ Khz).

QRY What is my turn?
Your turn is numbered ____.

QRZ Who is calling me?
You are being called by ___ (on ____ Khz).

QSA What is the strength of my signals (or those of ___)?
The strength of your signals is ___ (1. Scarcely perceptible; 2. Weak; 3. Fairly good; 4. Good; 5. Very good).

QSB Are my signals fading?
Your signals are fading.

QSD Are my signals defective?
Your signals are defective.

QSG Shall I send ____ messages at a time?
Send ____ messages at a time.

QSK Can you hear me between your signals, and if so, can I break in on your transmission?
I can hear you between my signals; break in on my transmission.

QSL Can you acknowledge receipt?
I am acknowledging receipt.

QSM Shall I repeat the last message that I sent you, or some previous message?
Repeat the last message that you sent me [message(s) number(s) ____].

QSN Did you hear me (or ____) on ____ Khz?
I did hear you (or ____) on ____ Khz.

QSO Can you communicate with ___ direct or by relay?
I can communicate with ___ direct (or by relay through ___).

QSP Will you relay to ___?
I will relay to ___.

QST General call preceding a message addressed to all Amateurs and ARRL members. This is in effect "CQ ARRL."

QSU Shall I send or reply on this frequency (or on ____ Khz)?
Send or reply on this frequency (or ____ Khz).

QSV Shall I send a series of Vs on this

frequency (or ___ Khz)?
Send a series of Vs on this frequency (or on ___ Khz).

QSW Will you send on this frequency (or on ____ Khz)?
I am going to send on this frequency (or on ____ Khz).

QSX Will you listen to ____ on ____ Khz?
I am listening to ____ on ____ Khz.

QSY Shall I change to transmission on another frequency?
Change to transmission on another frequency (or on ___ Khz).

QSZ Shall I send each word or group more than once?
Send each word or group two (or ___ times).

QTA Shall I cancel message number ____?
Cancel message number ____.

QTB Do you agree with my counting of words?
I do not agree with your counting of words. I will repeat the first letter or digit of each word or group.

QTC How many messages have you to send?
I have ____ messages for you (or for ____).

QTH What is your location?
My location is ___.

QTR What is the correct time?
The time is ___.

PROWORDS

Prowords are procedural words used to help ensure the accuracy and orderliness of communication between two or more stations. Examples include "wilco," which means that you have understood the message and "will comply" with its instructions. The proword "roger" is used to indicate that you have understood a message, though there is nothing saying whether or not you will comply with any instructions contained therein.

PROWORD	EXPLANATION
ALL AFTER	The remaining part of the message that follows the word _____.
ALL BEFORE	The part of the message that comes before the word ________.
AUTHENTICATE	The station is to reply to the challenge that follows:
AUTHENTICATION IS	The authentication on this message is _______.
BREAK	This separates one part of a message from the next.
CORRECT	You are correct, or what you have transmitted is correct.
CORRECTION	An error was made in transmission, what follows is the correct version (starting with the last word that was transmitted correctly).
DISREGARD THIS	This transmission is in error, disregard it.

PROWORD	EXPLANATION
TRANSMISSION-OUT	This proword shall not be used to cancel any message that has been completely transmitted and for which receipt or acknowledgment has been received.
DO NOT ANSWER	Stations called are not to answer this call. When this proword is employed the transmission shall be ended with "OUT."
EXEMPT	The stations following are not required to monitor this transmission.
FIGURES	Numerals or numbers follow.
FLASH	Precedence FLASH. Reserved for SHORT reports of initial enemy contact and emergency situations of vital proportion. Handling is as fast as humanly possible with an objective time of 10 minutes or less.
FROM	The originator of this message is as follows:
GROUPS	The message contains coded groups, the number of which follows:
I AUTHENTICATE	The transmission that follows is the reply to your challenge to authenticate.
INFO	The stations that follow are addressed for information.
I SAY AGAIN	I am repeating the transmission or portion requested.
I SPELL	I shall spell the next word phonetically.
I VERIFY	That which follows has been verified at your Request and is repeated. (Use only as reply to VERIFY.)
MESSAGE	A message that you are required to write down follows.
MORE TO FOLLOW	Transmitting station has additional traffic for the receiving station.
OUT	This is the end of my transmission, and no answer is required or expected. (Since OVER and OUT have opposite meanings, they are never used together)
OVER	This is the end of my transmission and a response is necessary. (Go ahead; transmit.)

PROWORD	EXPLANATION
PRIORITY	Precedence PRIORITY. Reserved for important messages that must have precedence over routine traffic. This is the highest precedence that normally may be assigned to a message of administrative nature.
RADIO CHECK	How well can you hear and understand my transmission?
READ BACK	Repeat this entire transmission back to me exactly as received.
RELAY (TO)	Transmit this message to all addressees. The address component is mandatory when this proword is used.
ROGER	I have received your last transmission satisfactorily.
SAY AGAIN	Repeat all or (ALL BEFORE___ or ALL AFTER___) of your last transmission.
SERVICE	The message that follows is a SERVICE message.
SILENCE	Cease transmission on this net immediately. Repeat three or more times.) The station imposing must authenticate.
SILENCE LIFTED	Silence is lifted. Authentication is also required.
SPEAK SLOWER	Your transmission is at too fast a speed.
THIS IS	This transmission is from _____.
TIME	That which immediately follows is the time or date-time group of the message.
TO	The addressees immediately following are addressed for action.
UNKNOWN STATION	The identity of the station with whom I am attempting to establish communication is unknown.
VERIFY	Verify the entire message (or portion) with the originator and send correct version (to be used only by addressee).
WAIT	I must pause for a few seconds.
WAIT OUT this frequency.)	I must pause longer than a few seconds. (Others may transmit on

PROWORD	EXPLANATION
WILCO	I have received your signal, understand it, and will comply. To be used only by the addressee. Since the meaning of ROGER is included in that of WILCO, the two prowords are *NEVER* used together.
WORD AFTER	The word in the message to which I have reference is that which follows ________________.
WORDS TWICE	Communication is difficult. Transmit (or I am transmitting) each phrase (or each code group) twice. This proword may be used as an order, as a request, or as information.
WRONG	Your last transmission was incorrect. The correct version is ________________.

LOGGING

There is no law that specifically requires that you maintain a radio log; however, there are many advantages to doing so. Keeping a radio log lets you keep track of your contacts, but more importantly, it lets you keep track of your communication capabilities.

If you find that you can make regular contact into an area on the 40-meter band but cannot contact the same area using the 20-meter band, you would know not to plan on using 20 meters for future contact into that area. If you have clear contacts during the day but poor contacts at night on any given frequency, again that is information that you will learn from studying your radio logs.

You can buy radio logs from most of the suppliers of radio equipment; however, I find that it is often better to customize log sheets to your specific needs. Simply create a master sheet and have copies made at your local print shop. The print shop can even bind these pages into a book for you. On page 52 is a sample of the log sheets I use.

COMMUNICATIONS ELECTRONIC OPERATING INSTRUCTIONS

A Communications Electronic Operating Instruction (CEOI) is the "how-to guide" for operating your communications system under your particular set of circumstances. The purpose of a CEOI is to standardize how everyone communicating within a given group will act under a particular set of circumstances.

Operating frequencies and times comprise a major portion of your CEOI. Communications will not be very effective if all stations are not on the same frequency at the same time. This is your complete PACE (Primary, Alternate, Contingency, Emergency) communications plan.

Other important considerations include lost communications procedures. For example, what do you do if you make contact on your primary frequency but lose contact before you have finished passing all of the traffic you have?

What are your missed communications procedures? If you are maintaining radio watch for a given station and that station fails to make contact when scheduled (i.e., PACE plan fails), what do you do? If you are the station that missed the scheduled contact, what should you do?

What circumstances dictate activating your emergency communications procedures? What are those procedures?

Your CEOI will also include your current authentication/cipher tables and very likely a copy of your brevity list.

Date	Time On /Off	Frequency/Mode	Station Contacted	Location
23 Aug 02	1835/1912	14.070 MHz/PSK31	KZZ9ZZZ	Los Angeles, CA

Power: 50 Watts
Antenna/Azimuth: G5RV
Signal Report (Sent/Received): 599 / 599
Weather: Clear/Sunny
Notes:

Date	Time On/Off	Frequency/Mode	Station Contacted/Heard	Location

Power:
Antenna/Azimuth:
Signal Report (Sent/Received):
Weather:
Notes:

Date	Time On/Off	Frequency/Mode	Station Contacted/Heard	Location

Power:
Antenna/Azimuth:
Signal Report (Sent/Received):
Weather:
Notes:

CEOI

TIME PERIOD 1 Page 1 of 10

Authentication/Cipher Table

TIME PERIOD 001 SERIAL NO. 001

	0	1	2	3	4	5	6	7	8	9
A	QJHK	CMR	STY	NE	IO	VBX	DL	WG	PU	ZF
B	QSEM	NIP	JVR	FU	LC	ZTX	AO	YD	KG	HW
C	HTBP	REA	NYF	OL	QV	KDU	IX	WG	JS	ZM
D	FWQO	GIT	KCJ	XR	MS	PYA	ZV	HL	UN	BE
E	RIQO	AYJ	XHM	GC	DZ	SNW	PF	TU	BL	KV
F	YXNH	LUQ	TRS	CG	ME	VDI	PB	WJ	KA	ZO
G	KUZC	LDY	WPJ	HM	AF	RSE	OI	QX	VT	NB
H	EYBN	OSF	TIW	UV	PJ	RZC	XD	AG	ML	QK
I	LRCT	KQW	MBG	VO	JD	FPX	NA	ES	UY	ZH
J	IZFC	TYR	MXG	LU	PD	SWA	EB	NO	KV	HQ
K	LTOY	HEA	SUX	GB	ZJ	NFI	PQ	MD	WV	RC
L	YTBU	FRD	HIS	EW	ZA	MOX	QN	VK	JC	PG
M	QASB	GYO	JZF	KE	VI	PWX	HD	CU	NR	LT
N	WOPQ	KUE	IFM	ZS	BV	AJH	LC	GY	TX	RD
O	XHTJ	GYA	IFN	SB	CU	WLQ	EV	PK	RZ	DM
P	FOKY	DZT	BVI	CX	GU	JWL	EA	MH	QR	SN
Q	BYGI	SHA	OZK	NL	WV	RCF	EJ	DM	XT	UP
R	SABJ	CGO	PHK	MT	WX	LQF	DI	ZE	YN	VU
S	FARM	XQP	KDJ	NT	EB	UYV	ZW	LI	CH	GO
T	DWCN	GOY	HAL	VR	MQ	IKE	ZX	FU	PB	JS
U	HWTX	YJD	KSV	FA	PG	OCZ	BE	NR	MQ	IL
V	FZLE	IMT	WBJ	XQ	DS	RPY	NC	GA	KH	UO
W	EOFZ	YIM	TCL	BX	NH	SRQ	DJ	GP	UK	VA
X	OEVR	WKS	HDL	FJ	AQ	PGY	TI	NZ	UC	MB
Y	LZKU	JOT	VXH	DC	NP	QGI	EB	AS	FW	MR
Z	LIMK	JFQ	RSE	OY	GH	UBV	NT	CX	DW	AP
^Set Letter										
A	BCDE	FGH	IJK	LM	NO	PQR	ST	UV	WX	YZ

Frequencies

Primary	7.235 (LSB)
Alternate	7.260 (LSB)
Contingency	10.140 (CW)
Emergency	10.125 (CW)
PACTOR (PBBS)	14.065
FM Repeater #1	147.250 *
FM Repeater #2	145.435 **
GMRS/Inter-Team	11/PL 22

* Repeater #1 PL Tone 103.5

** Repeater #2 PL Tone 100.0

Lost communications—Tune to contingency frequency to reestablish contact.

Frequency Chart

A frequency chart allows one to designate frequencies quickly and with a greater degree of efficiency under adverse band conditions. Whether transmitting by voice or by another mode (such as Morse code), it is easier to understand a three-letter group than it is a series of numbers in a given frequency.

Another advantage of having an established frequency chart is that it offers some degree of security in designating frequencies on the air. Telling everyone operating in a network to change to frequency ZYY will easily move the net, in our following example, to the 20-meter band/14.235 MHz. However, someone without access to the frequency chart will have to scan several bands and frequencies in an effort to follow the frequency change.

When creating a frequency chart, begin by choosing a number of frequencies in each band. In the following example we have chosen a few frequencies in each of the Amateur high-frequency bands. Next designate a three-letter group for each of these frequencies.

ZAA – 1.8100	ZQQ – 10.125	ZBH – 21.125
ZBB – 1.8300	ZRR – 10.130	ZBI – 21.325
ZCC – 1.8500	ZSS – 10.135	ZBJ – 21.375
ZDD – 1.9500	ZTT – 10.145	ZBK – 21.400
ZEE – 1.9855	ZUU – 14.035	ZBL – 21.425
ZFF – 3.5250	ZVV – 14.050	ZBM – 24.925
ZGG – 3.7000	ZWW – 14.100	ZBN – 24.950
ZHH – 3.8500	ZXX – 14.230	ZBO – 24.955
ZII – 3.8750	ZYY – 14.235	ZBP – 24.975
ZJJ – 3.8950	ZZZ – 14.250	ZBQ – 24.985
ZKK – 7.0500	ZBA – 14.325	ZBR – 28.500
ZLL – 7.1250	ZBC – 18.075	ZBS – 28.750
ZMM – 7.2250	ZBD – 18.115	ZBT – 29.000
ZNN – 7.2500	ZBE – 18.125	ZBU – 29.350
ZOO – 7.2750	ZBF – 18.140	ZBV – 29.650
ZPP – 10.110	ZBG – 18.155	ZBW – 29.675

The frequencies in the chart need not be distributed equally across all bands. Simply choose those frequencies that meet your communications needs and add them to your frequency chart.

MORSE CODE: A SURVIVAL SKILL

Morse code was invented by Samuel F.B. Morse and bears his name to this day. Initially used for telegraph communications and quickly adopted into the developing wireless (radio) service, Morse code provides an effective means of communication when conditions make other methods unreliable.

Compared to voice communication, Morse Code, or CW, uses a very narrow band width. This allows the signal to penetrate static and be effective at much lower power than is required for other types of communication. Furthermore, the dot, dash tones of Morse Code are more easily distinguished from background noise than is voice communication.

Originally Morse code was the primary means of communication on amateur radio, and on most other forms of radio as well. With advances in radio technology, voice communication and other modes, such as radio teletype, began to be added to the methods used by radio operators. However, Morse code still remained an important mode of communication, and perhaps served as the backbone of Amateur radio. Morse code is the only mode of communication permitted on *all* Amateur Radio frequencies, while band plans may limit the other modes of communication used on certain frequencies and bands.

In recent years, Morse code has taken on increasingly less importance as a radio communications mode, as other modes have been developed. The FCC reduced the Morse code speed requirement to five words per minute for all classes of Amateur Radio license, and services such as the U.S. Coast Guard have simply done away with the requirement that their radio operators be capable of sending and receiving Morse code.

Newer communication modes may well replace Morse code as the backbone of communications in various organizations and services. There might even be justification for doing away with the Morse code requirement all together as a qualifying skill for radio operators. However, when it comes to communications for survival and self-reliance, Morse code is still often the best—and perhaps at times the only—means of communication that will meet our needs. As an example of this, I have on occasion been using one of the "high-tech" computer-based modes of communication and had the computer lock up. This was not a major problem, since correcting the problem was simply a matter of rebooting the computer. However, to the receiving station I had simply disappeared. With Morse code capability, it was simple to just grab the code-key and send a quick message asking the receiving station to stand by while the computer rebooted. Because Morse code always works (assuming anything is going to work) I believe that it is a skill that should be learned and maintained by all survival-minded communicators.

Another primary advantage of Morse code is that one can provide accurate communication over great distance using only very basic equipment. Even the simplest radio kit can provide communication via Morse code over thousands of miles. The CW-only radios are small and operate at low power on batteries.

International Morse Code

Learning Morse code is not particularly difficult, but it does take consistency and practice. There are several Morse code training programs available on cassette tape, audio CD, and various computer programs. No matter which program or method you choose to learn Morse code, daily practice is the key to success.

Once you have learned Morse code, you will want to increase the speed at which you can send and receive. The idea is to learn Morse code as if it were a foreign language so that the different patterns of dits and dahs come to be automatically associated with letters (and later actual words). When you practice Morse code you should practice at a speed slightly faster than

International Morse Code

Letters

A	•—
B	—•••
C	—•—•
D	—••
E	•
F	••—•
G	——•
H	••••
I	••
J	•———
K	—•—
L	•—••
M	——
N	—•
O	———
P	•——•
Q	——•—
R	•—•
S	•••
T	—
U	••—
V	•••—
W	•——
X	—••—
Y	—•——
Z	——••

Emergency (SOS) •••———•••

Numbers

0	—————
1	•————
2	••———
3	•••——
4	••••—
5	•••••
6	—••••
7	——•••
8	———••
9	————•

Punctuation

Period	•—•—•—	(AAA)
Comma	——••——	(MIM)
Question Mark	••——••	(IMI)
Hyphen or Dash	—••••—	(DU)
Fraction Bar (/)	—••—•	(DN)

Symbols

Break	—•••—	(BT)
End Message	•—•—•	(AR)
Understood	•••—•	(SN)
Wait	•—•••	(AS)
Over	—•—	(K)
End of Work	•••—•—	(SK)

that which you can easily copy. For example if you find you can accurately copy at 5 words per minute, then practice at 7 words per minute. If you are comfortable at 7 words per minute, then practice at 10 words per minute. If you are comfortable at 20 words per minute, then practice at 25 words per minute.

Although it is nice to be able to send and receive Morse code rapidly (20 words per minute and faster), when it comes to Morse code it must be remembered that *accuracy transcends speed.* I would much prefer to communicate via Morse code at 10 words per minute with 95-percent accuracy than at 50 words per minute with only 50-percent accuracy.

Besides being the backbone of radio communication for survival and self-reliance, Morse code can also be sent by means other than radio. Blinking lights, a wig-wag flag, or anything that can be turned on and off or sent as long and short signals can be used to transmit Morse code.

CHAPTER 5

Digital Communications

Digital communications are text-based communications sent by connecting your computer to your radio and transmitting data from the computer across the airwaves. Although the concept of transmitting text data (nonvoice) via radio has been around since the beginning of radio itself, it is the advent of the personal computer that has made these modes popular. An entire book could be written on digital communications, but here we will look at the more common and popular modes and those modes that have specific application to survival and self-reliance communications. We will see that digital communications allow us to communicate under conditions that make other modes (such as voice) impossible.

SOUND CARD DIGITAL MODES

The widespread use of personal computers has led to the development of communications modes that integrate the computer and the radio. While using computers to aid in communication is not particularly new, the development of communication modes that use the computer's sound card to decode radio signals is a fairly new and unique addition to Ham radio.

It is possible to receive signals from these new sound card digital modes simply by connecting the audio out (i.e., headphone jack) on your radio to the audio in plug on your computer. Then just download the free software for the mode you are interested in using (or software to receive multiple modes) and watch the text of these messages appear on your computer screen.

Of course, to transmit you will need to have an interface between your radio and computer that allows the computer to key the radio and transmit the digital signal from the words you type on your keyboard. If you are skilled in electronics you can build an interface for just a few dollars. However, most people purchase a completed interface from companies such as Rigblaster, MFJ, or Rascal.

The interface works for all sound card digital modes, requiring that you install compatible software on your computer for only the mode you want to use. There are freeware programs available for all of the sound card digital modes. Simply search by mode name, using a standard Internet search engine to find download sources for these programs.

Once you have an interface and have

All that's needed to operate sound card digital modes from the backwoods: A laptop computer, a portable radio (in this case the Yaesu FT-817), an interface (Rigblaster Nomic), and an antenna (MP-1 Super Antenna).

downloaded the necessary software, you are ready to get on the air with these sound card/digital modes. These modes allow keyboard-to-keyboard communication under conditions where radio voice communication would be impossible. Furthermore, you can transfer files and data that would be impractical to transmit by voice.

A laptop computer makes sound card/digital communications a completely mobile/portable operation.

PSK31

PSK31 stands for "Phase Shift Keying 31-baud" and is probably the most popular of the sound card digital modes, allowing communication between computers connected to a radio.

PSK31 was designed by Peter Martinez (G3PLX) and is based on the Radio Teletype (RTTY) mode of operation. PSK31 is not intended to replace RTTY, but certainly makes an excellent addition to the communication modes available to Amateur Radio operators.

One of the major advantages of PSK31 is its narrow bandwidth. In this case bandwidth equals the baudrate of 31.25, giving us a bandwidth of 31.25 Hertz. This means that several signals can be (and often are) present on a single frequency. Furthermore, this narrow bandwidth allows PSK31 signals to get through using lower power and under conditions where voice communication would be impossible.

Once you have PSK31 set up and running you can engage in keyboard-to-keyboard chats with other PSK31 operators. In fact several different stations can participate in the same conversation, all stations monitoring the conversation and passing transmit time to each station in turn.

PSK31 sends and receives at about 50 words per minute. This is faster than all but the very best Morse code operators, and about the speed of a good typist. You can count on PSK31 to send pretty much at the speed you type.

If there is any disadvantage to PSK31 it is that it contains no forward error correcting function. This can lead to garbled text appearing on the screen as the PSK31 software interprets various static as signals and tries to display these signals. This is, however, a very minor problem. I find that most PSK31 contacts I have made

PSK31 Frequencies – Upper Side Band (USB)

1.838.15 MHz	
3.580.15 MHz	21.080 MHz
7.035.15 MHz (IARU Region 1 & 3)	24.920 MHz
7.080.15 MHz (IARU Region 2)	28.120 MHz
10.140 MHz	50.290 MHz
14.070 MHz	144.144 MHz
18.100 MHz	432.200 MHz

have near 100-percent good copy. As with any mode, if you are trying to work a very weak or distant station you might have less than perfect copy. However with PSK31 you will at least be able to work that station, whereas you wouldn't even know it was there if you were listening for a voice contact.

This having been said, there is in fact an error correction mode for PSK31. The mode is QPSK (Quadrature Phase Shift Keying), which adds a second carrier at a 90-degree phase difference to the PSK31 signal. Running QPSK will give you 100-percent clear copy on your received text, however, tuning is very critical when using the QPSK mode. Receiver accuracy must be less that four hertz to use QPSK effectively; otherwise it does not accurately detect the phase shifts. I run PSK31 in the QPSK mode at times and find that it helps clear up some signals, however, even without the QPSK mode PSK31 usually gives very readable signals.

I find that because of its ability to get weak signals through, PSK31 makes an excellent "travel mode," with any type of travel where you might take a laptop computer. With the addition of a portable HF radio (such as the outstanding Yaesu FT-817), an interface (such as the Rigblaster Nomic), and a portable antenna (such as the MP-1 Super Antenna), you have a highly portable PSK31 station. The entire system will easily fit inside of a briefcase or in the case in which you normally carry your laptop computer.

I have repeatedly made contacts using this exact set up with stations more than 1000 miles distant. In fact I have made frequent contacts between the east and west coasts of the United States using this setup, by switching to a G5RV antenna, while running between 2.5 and 5 watts output power.

If you have a computer and a radio, the sound card digital modes will greatly enhance your communications ability. It is also interesting to note that PSK31 works on the 11-meter CB radio frequencies just as well as it works on the HAM radio 12-meter and 10-meter bands. While FCC regulations prohibit nonvoice communication on CB radio frequencies in the United States, if you live in a country that allows nonvoice communication on CB radio, PSK31 would make an interesting addition to your CB station.

Once you have PSK31 up and running, you might want to begin using other sound card digital modes as well. The interface and setup is exactly the same for these other mode as for PSK31. You just need to download the software for the other modes.

Feld-Hellschreiber

Invented in 1929 by Dr. Rudolf Hell, the Hellschreiber system was designed for communication on wire systems (such as telegraph lines) but was quickly adopted for use in the radio service. In fact, several press agencies were using Hellschreiber to transmit their reports back to their home offices at the end of World War II.

With the popularity of sound card digital modes, software was developed allowing Hellschreiber signals to be decoded via computer and displayed on the computer screen.

Hellschreiber differs from PSK31 in that Hellschreiber is a "human readable" mode. Hellschreiber breaks letters into parts and transmits those parts to the screen, where they are in effect redrawn. This may be likened to the dots in the letters printed by a dot-matrix printer. Each dot is sent separately, with on/off keying printing the dots and blanks to form the letters on the screen.

Hellschreiber is an excellent mode for use under poor transmission/reception conditions. I have found that even when conditions prohibit PSK31 signals from being received, it is still often possible to communicate using Hellschreiber.

Transmission speed using Hellschreiber is about 25 words per minute. Although Hellschreiber signals are authorized on any frequency allocated to CW transmissions, the common operating frequencies for Hellschreiber

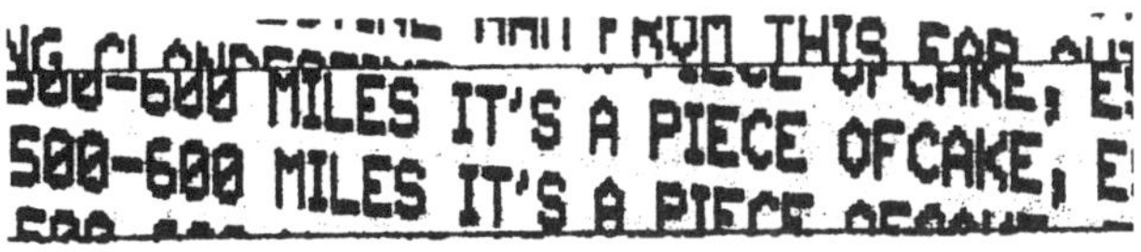

An example of Hellschreiber text.

are located around the 7.030 MHz, 14.063 MHz, 18.104 MHz, 21.063 MHz, and 28.063 MHz frequencies frequencies.

There are many people working on the development of Hellschreiber as a mode for Amateur Radio, but special recognition must go to Antonino Porcino (IZ8BLY) in Italy, and Murray Greenman (ZL1BPU) in New Zealand as the leading developers/advocates for this mode.

In addition to the standard Hellschreiber mode, the adoption of Hellschreiber into Amateur Radio has lead to the development of additional Hellschreiber modes. Among these modes are the following:

Duplo-Hell—A system using two tones to transmit the Hellschreiber signals. Duplo-Hell is a good mode for use on noisy frequencies, allowing signals to be received through the noise.

FM-Hell—A very narrow band signal. FM-Hell is not affected by flutter, and serves well for low-power, long-distance communication.

PSK-Hell—Very similar to FM-Hell, PSK-Hell is a narrow band signal serving well for low-power, long-distance communications.

Although perhaps not quite as popular as PSK-31, Hellschreiber is an active mode in use by many Amateur Radio operators. Because it is human-readable (i.e., you see the letters drawn on the screen and determine what they are), Hellschreiber may be able to get signals through, where other modes fail. As a means of communication for survival and self-reliance, Hellschreiber is a robust mode that should be in your communications package.

MFSK16

MFSK16 is a multi-tone frequency shift key mode using 16 tones to transmit signals. These 16 tones are sent at 15.625 baud and are 15.625 hertz apart. The overall bandwidth of a MFSK16 signal is only 316 hertz, providing good transmission quality for weak signals and under conditions of poor signal conditions. Although this bandwidth is wider than some of the other sound card digital modes, it is still narrow enough to be easily used in conjunction with the common 500 hertz CW filters.

One of the major advantages of MFSK16 is that it incorporates forward error correcting (FEC), thereby providing greater accuracy of received transmissions. This allows for a 31 word-per-minute transmission rate with the FEC function on, or a 61 word-per-minute transmission rate with FEC turned off.

While MFSK16 is a good mode for long-distance and weak-signal communication, it has not developed the popularity and following of modes such as PSK31 or even Hellschreiber. A March 2001 survey on the ARRL Web site indicated that only 13.9 percent of those responding had ever tried MFSK16, and 39.6 percent did not even know of what MFSK16 is. This does not make MFSK16 an ideal candidate for general calling just to see what contacts can be made. It does, however, make for a good mode of communication for scheduled contacts, offering a small degree of privacy due to the lack of general usage and limited popularity of this mode.

MT63

MT63 is a very robust mode capable of getting signals through under adverse band conditions. MT63 is intended for keyboard-to-keyboard communication and is specifically designed to work under poor signal conditions.

MT63 uses 64 tones spaced 15.625 Hz apart, thus making a MT63 signal one kHz wide. It also employs FEC (forward error correcting) so that MT63 is capable of printing 100-percent readable text, even if conditions distort up to 25 percent of the actual transmission. Each letter transmitted via MT63 is spread over several tones to reduce interference from other stations and atmospheric noise.

Throughput with MT63 is about 100 words-per-minute, which is faster than most people can type anyway (or at least faster than I can type). Overall this makes MT63 an excellent candidate for keyboard-to-keyboard communication. I personally find MT63 better than standard RTTY, and even better than PACTOR under adverse conditions.

Another advantage of MT63 is that it is a

somewhat obscure mode. It is freely available to anyone who wants to download a copy off the Internet and use it, but most people simply don't use MT63 (many even being unaware of its existence). This means that when communicating with MT63 there is a much smaller chance of having your traffic monitored than there is when using a highly popular mode such as PSK31. Along this same line, MT63 has a rushing sound (similar to background noise) when heard on the air. This means that someone hearing an MT63 signal on any given frequency could mistake it for simple static and not recognize that it is a form of communication.

The only disadvantage I see to MT63 is that it requires a fairly fast computer to run the MT63 software properly. You will need at least a 166 MHz speed computer to run MT63. This is generally not a major problem if you are using a new computer in your communications system, but if you use an older computer, MT63 will not function properly at slower processing speeds.

As long as you have a computer capable of running MT63 software (166 MHz or better) I recommend MT63 as a sound card digital mode for your survival and self-reliance communications package.

EchoLin2k

EchoLink www.echolink.org/el/) is a freeware program that allows HAM radio operators to connect their radios to the Internet and communicate worldwide using voice-over TCP/IP protocol. EchoLink allows radio-to-radio, computer-to-radio, and computer-to-computer connections.

To connect your radio to your computer, you will require a sound card interface. The good news is that if you are set up with PSK31 and other digital sound card modes, using an interface such as RigBlaster, you already have the interface needed for EchoLink. Simply download the EchoLink software and run it on your computer in the same manner as you do when using other digital/sound card programs, such as PSK31.

The first time you access the EchoLink servers, EchoLink will verify your Amateur Radio call sign before you are able to connect to other stations. Once your call sign has been verified, you will see a list of the various stations connected through EchoLink each time you log in. Because EchoLink is a two,-way system, you must have an Amateur Radio call sign to access the system. There is no provision for accessing the system on a "listen only" basis.

After your call sign has been verified, you connect to the EchoLink servers through your normal Internet connection. If you have a microphone on your computer, you can use this to talk from your computer to any other station/computer on the EchoLink. Two EchoLink users could set up computer-to-computer voice communication from any two places on earth having Internet access. Of course this requires that you have access to a computer and Internet at your given location. However, the advantage and intent of EchoLink is that you can extend your communication capability by adding radio to this setup.

With your radio connected to your computer via a sound card interface (e.g., RigBlaster) you can talk through your radio to your computer. That signal then travels across the Internet through the EchoLink servers and on to any other user linking to your call sign with the EchoLink software. Of course your radio is going to have to be sitting beside your computer in order to make the connection between the computer and radio. After all, the cables between your radio and computer are only so long. But once you have this connection made, you can use any other radio to access this connection.

As an example of an EchoLink setup, I have a 2-meter radio connected to my computer, using a RigBlaster interface. The radio is connected to an external antenna mounted atop a tall mast. Using just the radio, I can communicate over a fairly extensive local area talking with mobile and handheld radios operating on the same 2-meter frequency I am using. Now I bring up my EchoLink program, and any of those mobile or handheld radios that I can talk with from the radio can now send their radio signal to any place on earth, using the Internet voice-=over TCP/IP protocol. If you have ever listened to a commercial radio station on the Internet, the concept is the same, except as a HAM radio operator you have two-way communication!

By leaving my radio on and set to a particular frequency, and by having the EchoLink software running on my computer, I can take my handheld 2-meter radio and literally talk around the world. There is excellent potential for using this setup for emergency and rescue operations.

TNC DIGITAL COMMUNICATIONS

Having looked at the various sound card digital modes, we move on to modes requiring a terminal node controller (TNC) radio modem. These radio modems connect between your computer and your radio, allowing data to be transferred from the airwaves to your computer screen.

There are various manufacturers of TNC radio modems. Some of these manufacturers are:

Special Communications Systems PTC-II and PTC-II e modems – www.scs-ptc.com
HAL Communications – www.halcomm.com
Kantronics KAM Series modems – www.kantronics.com
Timewave Technology Series modems – www.timewave.com

Which TNC radio modem is best is pretty much a matter of personal preference. Once you have chosen your TNC modem, it's a simple matter to set up your radio mailbox using PACTOR or similar protocols.

PACTOR

PACTOR is a communications mode developed by a group of HAM radio operators in Germany in the1980s. PACTOR allows reliable communication at 100 to 200 baud, with a throughput of up to 18 characters per second.

One of the greatest advantages of PACTOR from a "survival communications" viewpoint is the ability to send and receive non-real-time (NRT) messages with another station. This is accomplished through something called a packet bulletin board system (PBBS). A PBBS is essentially a private mailbox that you can access via radio with a TNC modem.

You prepare a message in much the same manner as you might for Internet e-mail. However, instead of using your telephone to dial an Internet Service Provider (ISP), you use your radio to contact a PBBS on a given frequency. Once you have established a successful connection, your message can be uploaded to the PBBS. The message you upload may, of course, be addressed to the owner of the PBBS but may also be addressed to any other person or station that can access the PBBS.

An excellent overview of PACTOR and its use with the WinLink network is Jim Corenman's (KE6RK) A PACTOR Primer, available online at www.airmail2000.com/pprimer.htm.

WinLink

WinLink (http://winlink.org) is a radio/e-mail network allowing Amateur Radio operators to send and receive e-mail from their radios both to other WinLink users and to any valid e-mail address on the Internet. The WinLink user, whether on a ship at sea, traveling in a vehicle, or hiking in the wilderness, has the ability to send and receive e-mail from friends and family at home, even when in the most remote areas.

To use WinLink, you will need a radio capable of transmitting and receiving on the Amateur Radio (HAM) bands, a TNC radio modem, and a computer to prepare and read e-mail.

WinLink is accessed using the PACTOR protocol. You will also need to download a copy of the AirMail computer software needed to format your WinLink messages. This software is free and is available from the AirMail Web site at www.airmail2000.com. If you use Internet e-mail with the associated browser, AirMail software will be familiar. It also comes with a detailed "help file" that explains its functions.

Once you have PACTOR capability and have downloaded your AirMail software, you will need to register as a WinLink user. You can do this by simply logging into any of the WinLink stations. To log into a WinLink station you must ensure that your computer is connected to your TNC (radio modem) and that the TNC is connected to your radio. Set your radio to one of the WinLink frequencies and connect, over the air, to the WinLink station of your choice.

Once you have logged into a WinLink station, you become a registered user and are assigned a WinLink e-mail address—which is your-callsign@WinLink.org. Thereafter, any Internet user can send you e-mail by sending a normal e-mail to your WinLink e-mail address. You can send e-mail to any Internet e-mail user in the same manner.

Although WinLink works much the same as Internet e-mail, there are a few important differences of which you should be aware. First, there is no privacy in any message sent via WinLink. Any person with a radio, TNC radio modem, and the appropriate software can copy and read messages sent through the WinLink system. Because WinLink uses Amateur Radio, it is governed by the rules and regulations governing Amateur Radio operators in the country in which they are licensed. Use of Amateur Radio to enhance one's pecuniary interest is universally prohibited. Thus WinLink messages should not contain "business information." Furthermore, all messages sent across the WinLink system must be in plain language or in a format read with commonly available software. Amateur Radio regulations, and thus WinLink, specifically prohibit using encryption to obscure the meaning of messages sent on the Amateur Radio bands.

This being said, WinLink is an absolutely outstanding service. There are WinLink stations throughout the world, allowing you to gain access to WinLink from almost anywhere.

I have personally accessed WinLink while camping on a mountainside in the Pacific Northwest. Using a Yaesu FT-817 radio, Kantronics KAM-XL modem, and a laptop computer, I was able to exchange e-mail with friends at home using their Internet e-mail accounts. While a laptop computer may not be something that one would normally include on a backpacking trip, it certainly isn't an impossibility. In the case of my camping trip, I packed along the laptop computer and radio gear specifically to experiment with WinLink from a wilderness location. If, however, you are traveling by automobile or on a trip where your laptop computer is needed anyway, the addition of a small HF radio and a TNC radio modem puts you on the air with WinLink!

The map shows the current WinLink stations at the time of this writing. These stations are linked together in a network configuration via the Internet. This serves to coordinate e-mail and position reporting within the WinLink

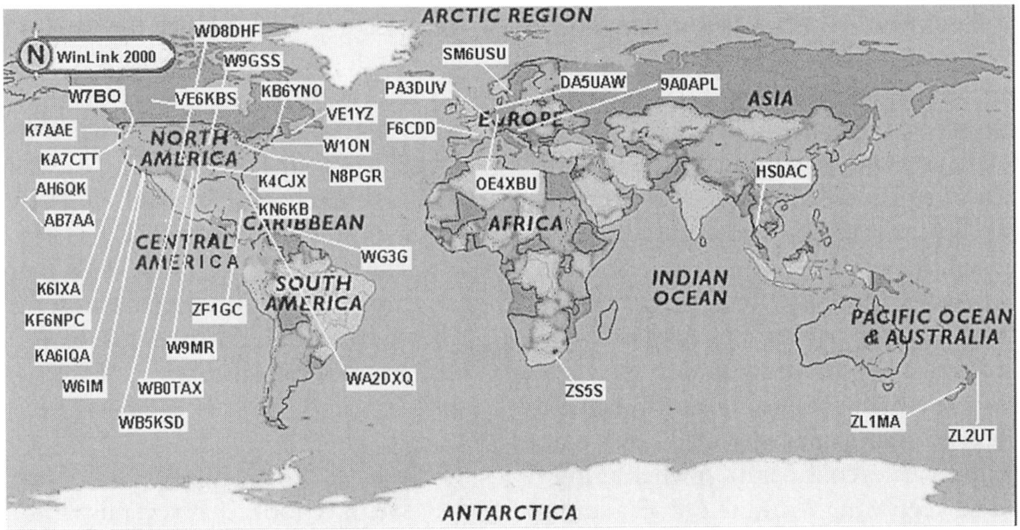

WinLink 2000 Map. (Used with permission of Steve Waterman, K4CJX, and the WinLink Development Team.)

system. Furthermore, the Internet backbone of the WinLink system leaves the radio frequencies free for Amateur Radio operators to use when accessing WinLink.

Each WinLink station operates on several different frequencies. This allows Amateur Radio operators to choose the best frequency/band based on the current signal propagation from their location to a WinLink station. The WinLink frequencies are contained in the AirMail software used to access WinLink and are available on the WinLink Web site.

Because of its networked configuration, you no longer need to choose a WinLink home station to log into. Once you log into any WinLink station, the system remembers this and makes your WinLink e-mail available from that station, as well as any other WinLink station you have logged into within the past 90 days. E-mail received at your WinLink e-mail address will be forwarded to all WinLink stations where you have connected within the past 90 days. After you retrieve your e-mail from any WinLink station, a message is sent across the WinLink network causing that message to be deleted from the other WinLink stations holding it for you.

If you use any of the common Internet e-mail services, you will have no difficulty using WinLink. However, there is some difference between e-mail on the Internet and e-mail sent and received via WinLink. The primary difference is transmission speed. WinLink is much slower than your dial-up connection at home or in the office. This means that you should try to keep your messages short when sending them via WinLink, and you should request that anyone sending e-mail to your WinLink address from the Internet likewise try to keep messages to a reasonable length.

You are also generally limited to the amount of access time you can have to the WinLink system. This is usually around 60 minutes per day. This is more than enough for most people to send and receive their personal e-mail every day.

A very important consideration when using WinLink is to keep your WinLink e-mail address semiprivate. You will of course provide it to friends and family with whom you want to communicate when you are in some remote location, or on the road. However, if your WinLink address falls into the hands of the criminals who flood your Internet e-mail address with unsolicited commercial e-mail (SPAM), it will become unusable almost instantly. Downloading the flood of illegal drug offers, child pornography, and get-rich-quick scams sent by spammers will quickly burn up your daily WinLink access time (not to mention violating the terms of your Amateur Radio Service license, which prohibits the use of Amateur Radio for commercial purposes). The WinLink administrators do a great job of keeping spam off the WinLink system, but it is essential that all WinLink users take steps to protect their address from criminal spammers.

G-TOR

G-TOR is a system designed by M. Golay and used by the Voyager spacecraft to send data back to Earth from Saturn and Jupiter. G-TOR stands for Golay Teleprinting Over Radio. G-TOR is a proprietary mode of the Kantronics corporation and is incorporated into its KAM series TNC modems. It is incorporated into just about every KAM modem sold, and there are thousands of KAM modems in the HAM radio community.

G-TOR is very stable communications mode, performing very well under adverse signal conditions. Furthermore G-TOR is capable of transferring data at twice the rate of PACTOR, and under adverse conditions is even comparable to PACTOR II.

I like G-TOR and use it when transmitting large files to another G-TOR-capable station. Unfortunately, even though G-TOR offers outstanding performance, it has not caught on in the same way as PACTOR. This is not a major concern if all stations agree to use G-TOR (and thus purchase KAM modems) as a communications means.

* * * * *

We have looked at several (although certainly not all) of the digital modes available for use with your radios. These digital modes range from the very popular PSK31 to the less common MT63 to the current backbone of digital communications, PACTOR.

Any one of these digital modes can provide the basis for a useable communications network. By taking advantage of the best features of a combination of these (e.g., PSK31 for chat and general conversation, PACTOR for radio e-mail and PBBS functions, and MT63 for communications under poor signal conditions), one can have an excellent communications network, useable under any survival or self-reliance situation.

I recommend digital mode radio communications for anyone setting up a personal communication system. Requiring only a radio, interface, and computer, digital mode radio communications can be run from almost any type of base station. With a laptop computer, digital modes can be run from any mobile or field site. I have personally run coast-to-coast communications using PSK31 and operating from a field site. These digital modes provide the most reliable means of radio communications, under the most varied conditions, of all modes (with the possible exception of Morse code).

For self-reliance communications (that is, communications where YOU control the system without reliance on telephone lines, wireless network companies, and the like), digital mode radio communications are almost the perfect answer. Useable anywhere from a penthouse in the largest city to a tent on the most remote mountainside, to a ship at sea, digital mode radio communications provide accurate, rapid, and reliable communications.

CHAPTER 6

Field and Mobile Communications

When choosing a radio for survival and self-reliance communications, there are certain things that you must consider. Of course, whatever radio you happen to have is better than any radio that you don't have, but when considering purchase of a new radio, some choices are better than others.

My three favorite radios for survival and self-reliance communications/field use: the SGC SG2020, the Yaesu FT-897, and the Yaesu FT-817.

I personally prefer Yaesu radios because they offer the features I find most useful.

The SG-2020 is an HF SSB radio that is built extremely tough. This radio operates on high-frequency SSB modes only. It also has the advantage of filtering out HF broadcast (shortwave) stations for those who do not wish to listen to them. The SG-2020 is lightweight, easily portable, and excellent for field use.

The Yaesu FT-897 is the high-power version of the FT-817. The FT-897 is pictured with the optional FC-30 antenna tuner attached. This is Yaesu's latest addition to the field radio market. The FT-897 is an outstanding radio, with its own internal power supply (two battery packs). The radio operates on all Amateur Radio HF bands as well as on the 2-meter and 440 VHF/UHF bands. The FT-897 is capable of 100-watt power output when run off an external 13.8 VDC power supply. Using its internal batteries, it has a power output of 20 watts.

The Yaesu FT-817 is perhaps the best field radio available today. It is lightweight and capable of multiband operation. Like its big brother (the FT-897), the FT-817 is capable of operating on all Amateur HF bands as well as on the 2-meter and 440 VHF/UHF bands. The

The author operating radios from the field.

FT-817 is capable of 5 watts power output across all bands. The FT-817 is easily carried in a belt-pouch or stowed with one's outdoor gear.

When it comes to handheld (walkie-talkie) radios, I again choose Yaesu for my field gear. The VX-5 and the waterproof VX-7 are radios designed with the outdoor adventurer in mind. Both radios operate on the 6-meter, 2-meter, and 440 Amateur Radio bands. The U.S. version of the VX-7 also has 300 mW transmit capability in the Amateur 220 band. Both the VX-5 and VX-7 include access to the NOAA weather band and have the option of installing a barometric sensor, thereby allowing the radio to display barometric pressure and altitude. These radios include DTMF (Dual Tone Multi-Frequency, commonly known as Touch Tone) features allowing you to access auto-patch features of area repeaters and make telephone calls from your radio.

Another major advantage of these excellent radios is their wide-band receive capability. You can monitor (receive, but not transmit) local broadcast radio stations. This allows you to listen to local news broadcasts and even enjoy the entertainment of music and talk radio programs.

For pocket-size VHF/UHF communication, either the VX-5 or the VX-7 is an excellent choice.

Although we have looked at my personal choices for field radios, these radios may not be the best for your particular application.

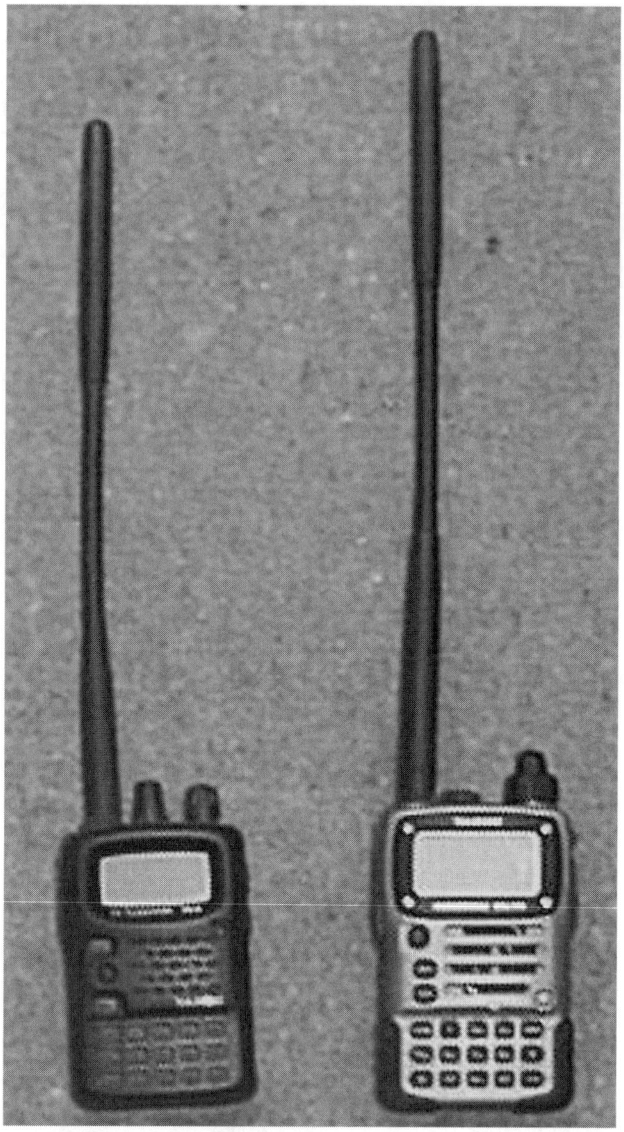

Yaesu VX-5 and Yaesu VX-7.

Any of the radios produced by a major radio manufacturer (e.g., Icom, Kenwood, Yaesu, Ten-Tec, SGC, Alinco) are likely to be of good quality.

MOBILE RADIO COMMUNICATIONS

I am strongly in favor of portable (man-pack) radios and communication systems, but there is another aspect of portable radio systems that should be considered—that is, mobile (vehicle mounted) radios.

My friend Radio Ray, a HAM radio operator

and communications expert in his own right, favors mobile radio systems for survival communications. There are some good advantages to having radios installed in your vehicle, regardless of what other communications systems you have in place.

The primary advantage of mobile communications is that they are mobile. Generally, whenever we travel away from our homes, we do so by automobile. In case of an emergency or other circumstance that forces us to leave our homes, it is likely that we will, at least initially, travel by automobile. Having radios installed in your vehicle means that you have communications as long as you are with your vehicle.

Your automobile can also serve as a temporary shelter. In a survival situation where you were stranded and could not return home, your vehicle could provide you with shelter from the elements. Sleeping in your vehicle overnight might not be the most comfortable option, but if a winter storm blocks the roads and you can't make it home, your vehicle will keep you safe until the roads are cleared. If you have radios installed in your vehicle, you can let friends or family know that you are stranded but that you are OK and are just waiting for the roads to be cleared.

Many radios that you would run as a base station can easily be run as a mobile station. Your vehicle already has a 12-volt battery that is constantly recharged when the motor is running. As long as your radios are designed to run on 12-volt DC power (as opposed to 110-volt AC) it's a simple matter to connect the radio's power cable to your vehicle battery—and you have a working radio. One important consideration here is to ensure that your vehicle alternator can handle the power draw of your radio. While the radio itself will probably not be any problem, some people install amplifiers for their radios. A high amp draw from an amplifier can be too much for the alternator of a small vehicle to handle.

Radios can be installed in many ways, depending on your vehicle configuration and the type and number of radios you are trying to install. One of the best and most secure ways of mounting radios in your vehicle is to install them as an in-dash mount.

You likely already have an AM/FM type radio mounted in your vehicle. This radio can be replaced with a HAM radio. Since many HAM radios have wide band receive capability, you will still be able to tune in your favorite music station or listen to your favorite talk-radio program on your way to work each morning. However, having a HAM radio installed in your vehicle means that you can also switch to any of the authorized Amateur Radio frequencies and have two-way communication to almost anywhere in the world.

In addition to an HF radio for worldwide communications, you might also want to install a UHF/VHF radio (e.g., 2 meter/440) or GMRS radio for local communications. Local FM communication (and even regional communication through the use of repeaters) is very useful during a local emergency, as well as for more mundane uses, such as avoiding traffic jams.

You will also need to mount antennas on

Amateur radios mounted in vehicle.

your vehicle. However, this need not be a high-profile affair. A couple of antennas mounted on your vehicle will attract little if any attention.

If you are running both HF and UHF/VHF radios, you will need at least two antennas. Your HF antenna should be connected to an automatic antenna tuner, thereby allowing you to work across the entire Amateur Radio HF frequency allocation.

The antenna for your UHF/VHF radio is generally matched to the band in which you are operating and will not need an antenna tuner. The frequency difference from one end of a given UHF/VHF band (e.g., 2-meter band 144.000 MHz – 148.000 MHz) is not so great as to require retuning of the antenna from one frequency to the next within the band.

The basic concept is a power source (your vehicle battery), a filter to reduce electric noise from the vehicle itself, the radio(s), an antenna tuner (for HF radios), and a good-quality antenna.

Once you have your mobile system installed, you can conduct almost any type of communication from your vehicle that you would conduct from a base station. The main difference you will find in mobile communications is not in the radio. In fact, the same radios used for mobile communications can be (and often are) installed as base stations. The main difference is that mobile antennas must, as a matter of necessity, be smaller than base station antennas, and, of course, they cannot be mounted nearly as high above the ground.

ELTs AND EPIRBs

ELT and EPIRB stand for Emergency Locator Transmitters and Emergency Position Indicating Radio Beacons, respectively. As the names indicate, these devices are radio beacons

Vehicle-mounted antennas.

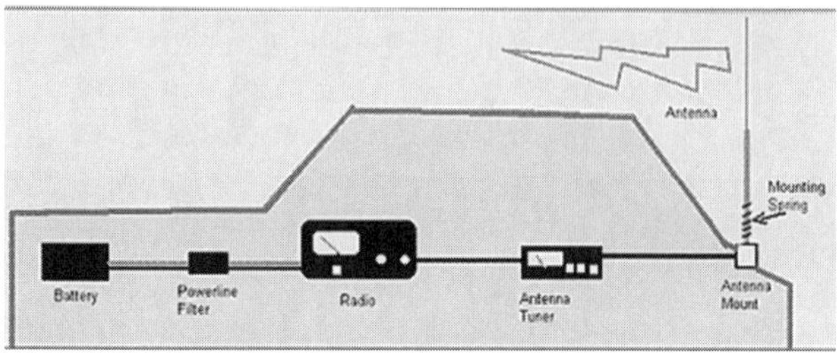

Schematic of vehicle-mounted radio system.

Close-up of antenna tuner mounted on vehicle.

intended to signal distress and the location of the person/vessel/aircraft in distress.

These transmitters/beacons can be divided into three broad categories: ELT homing beacons operating on 121.5 MHz, COSPAS/SARSAT beacons operating on 406 MHz, and INMARSAT beacons operating on the 1.6 GHz L-band.

The ELT transmits a sweeping tone on 121.5 MHz (or 243.0 MHz military) but does not actually transmit its position. These signals can be received by satellite, but the satellite does not determine the position of the transmitter. Upon activation of an ELT transmitter, all the satellite knows is that there is an active ELT somewhere in its footprint below. The position of the ELT can be narrowed down somewhat by measuring the Doppler shift of the signal, but even a small error in the frequency can result in large errors in pinpointing the position of the ELT.

Rescue teams locate the ELT by triangulating its position using radio direction finding techniques. Many search and rescue (SAR) aircraft carry radio direction-finding equipment, as do some ground teams. However, since most ELT activations are false alarms, an ELT signal does not result in the immediate deployment of SAR teams.

EPIRBs operating in the 406 MHz range have several advantages over the older ELTs. First is the stability of the signal on 406 MHz, allowing greater accuracy in pinpointing the location of the EPIRB using Doppler shift. The satellite is usually able to determine the location of the EPIRB to within several miles using Doppler shift. It then stores this information and downloads it to an earth station when its orbit takes it within range. Due to the use of low earth orbiting satellites, it may take from 15 minutes to 4 hours for the position information to be relayed to an earth station, depending on the orbital path of the satellite.

Another advantage of the EPIRBs is that they can be programmed with the vessel's name, Maritime Mobile Service Identity (MMSI), or other identifying information that is transmitted with the distress signal so that SAR knows who is sending the signal.

The 406 MHz EPIRBs also transmit a 250 mW 121.5 MHz homing beacon. This is because most rescue helicopters are equipped with radio direction-finding equipment specifically designed to home in on the 121.5 MHz beacons as they near the rescue site.

Finally, EPIRBs can be equipped with a Global Positioning System (GPS) receiver so that an exact position can be determined from the GPS and transmitted with the EPIRB signal. Thus, the EPIRB sends a distress signal, identifies who is in distress, and gives a precise location of the person in distress.

The latest type of EPIRB is the INMARSAT EPIRB operating in the 1.6 GHz L-band. The INMARSAT satellites are in geostationary orbits, meaning that they hold their position above the earth. The INMARSAT EPIRB contains both GPS reporting and the 121.5 MHz homing beacon.

GLOBAL POSITIONING SYSTEMS (GPS)

The Global Positioning System was developed by the Department of Defense and consists of 24 satellites in orbit at 20,200 kilometers above the earth, inclined on an angle of 55 degrees. The satellites are arranged in six orbital planes with four operational satellites in each plane. The last of these satellites was put into place on March 9, 1994, and since that time there has been consistent worldwide GPS satellite coverage.

A GPS receiver works by receiving signals from these satellites and using triangulation to determine a position. Each satellite transmits two signals—a coarse acquisition code, intended for civilian use, and a precision code, intended for military/government use. The coarse acquisition code provides position accuracy to within about 100 meters, while the military precision code provides accuracy to within less than 20 meters. The coarse acquisition code is intentional but can, in fact, be turned on and off, depending on the whim of the GPS Master Control Station located at Falcon Air Force Base near Colorado Springs, Colorado.

It is important to be aware that your GPS

receiver might have an error in precision of up to 100 meters, based on reception of either the coarse acquisition code or the military precision code. This normally isn't any great problem, since even given worst case, your GPS will get you to within about 100 meters of where you are going (and usually much, much closer).

Transmitting Coordinates

Giving your location is an important and common skill in survival communications. Unfortunately, errors in transmitting coordinates are common. The USAF Pararescue manual states,

> Many inaccuracies may exist when comparing an actual site location with its identified map location. These inaccuracies come from plotting/pulling errors, datum transformation errors, symbol displacement errors, and incorrect specifications. These errors can be compounded if datums are mixed when reporting positional information. When passing or transmitting coordinates, it is important that the complete source of the coordinates be given. In order to avoid confusion, the source will include the map or chart producer, series, sheet number, edition, date, and datum.

It is important to practice with your GPS, to practice map-reading skills, and to further practice transmitting accurate locations to others. Try going to a specific location and determining the coordinates for that location with your GPS. Transmit those coordinates to a friend who programs them into his own GPS and navigates to your location. Once your friend locates you, he moves to a new location and sends a GPS location to you, allowing you to navigate to him. Such games make for an interesting afternoon outing, while developing skill in the use of your GPS and radio gear. Once you have developed some skill with your GPS and navigating, add new challenges to the game. Try sending the GPS coordinates using some type of code or cipher. You and your friend will now need to decrypt the message in order to program the coordinates into your respective GPS units.

GeoCaching

Another very enjoyable way of developing skill with your GPS is to join the international game of geocaching. To play, you hide a small cache containing trinkets or souvenir items and determine the coordinates for your cache using your GPS. Thereafter, you post your cache description and coordinates to the geocaching Web site at www.geocaching.com.

You can use your GPS to locate the caches of others. In each cache is a logbook where you can leave a brief message and the date and time you located the cache. You then leave a small trinket and take one from the cache. Thus, the contents of the caches are constantly changing.

Geocaching is a fun outdoor activity that works to build navigation skills with your GPS. Visit the geocaching Web site and sign up to play. It's fun, it's free, and it improves your survival skills.

SELF-POWERED RADIOS

If I could have one and only one radio for survival and self-reliance, it would be one of the AM/FW/SW self-powered radios. The major advantage of these radios is that they require no external power source and no batteries! To power your "self-powered" radio, you simply wind it up. By turning a crank on the outside of your radio, you power a dynamo and charge internal power cells, thereby giving power to your radio.

Arguably the best self-powered radio currently available is the Freeplay Plus, sold by the C. Crane Company. In addition to the hand crank, the radio also contains a solar panel to charge its internal power cells. It can also run off of AC power (which charges the internal power cells) when commercial power is available.

As an added and useful feature, a detachable LED light on a retractable six-foot cord is powered from the radio. This provides a small amount of light while using the radio.

For greater light, FreePlay makes a self-powered light, which also serves as a 3-volt generator capable of powering small radios. While the FreePlay Plus is the best radio overall,

Free-Play Plus radio. (Photos courtesy of C. Crane Company.)

FreePlay Light and Generator, FreePlay Plus Radio, and Radio Shack Self-Powered Radio.

it is slightly large to be conveniently portable in a rucksack or similar pack. For portable use, the Radio Shack self-powered AM/FM/SW/WX radio is an excellent choice.

The Radio Shack self-powered radio runs off of its internal power cell, which is charged with the external crank or can be powered by two AA batteries. It is my second choice for a self-powered radio. Another advantage of the Radio Shack radio, especially as a portable radio, is the inclusion of the NOAA Weather Band, allowing quick access to local weather forecasts.

No matter which one you choose, I believe that a self-powered radio is an essential part of your survival and self-reliance communications planning.

SCANNERS

In simple terms, a scanner is a radio receiver that monitors several preprogrammed frequencies in rapid succession, stopping on active frequencies to allow the radio traffic to be heard.

Most often when we think of scanners, we think of them in terms of monitoring police and fire frequencies. The ability to monitor police and fire radio traffic, as well as that of other public service agencies, makes radio traffic a valuable asset for survival communications planning. This kind of monitoring gives us

Uniden Bearcat BC895XLT and RCA ScanTrak RP-6150.

firsthand information about situations in our area that might affect us. Hearing the highway patrol respond to an accident blocking a certain road gives us a good indication that we should take an alternate route to avoid the resultant slowdown or traffic jam. Hearing the police respond to a "man with a gun" call at the local shopping mall provides a good indication that we might want to postpone our shopping trip for a little while.

The advantage of having a scanner capable of monitoring police, fire, and other public service frequencies in your home is multiplied greatly when that scanner is mounted in your vehicle. It is much more important to be aware of potential hazards and thus avoid them while actually traveling than when just planning the trip.

Unfortunately, a few states have passed laws that proscribe the use of a scanner in a vehicle. These states seem to believe that, while government agencies and officers may use radios, scanners, and the like, citizens have no right to do so. The states that seem to have the most restrictive/oppressive laws regarding mobile radio scanners are Florida, Indiana, Kentucky, Michigan, Minnesota, and New York.

Even in states such as California, where the state law simply prohibits use of a scanner in the furtherance of a crime (and a few other states have similar laws), there might be local or county laws that prohibit mobile radio scanners. For example, the following is a Los Angeles County ordinance:

> 13.10.020 Installing or using shortwave radios in vehicles prohibited without permit. Except as provided in Section 13.20.050 of this chapter, every person who, without obtaining a permit from the sheriff or from the forester and fire warden authorizing him to do so, equips any vehicle with, or operates any vehicle equipped with, a shortwave radio receiver, is guilty of an infraction. (Ord. 83-0066 § 90, 1983: Ord. 5462 § 2, 1950: Ord. 4322 § 2, 1944.)

The following are examples of state laws where scanners are generally prohibited.

Florida Scanner Law

Title XLVI - CRIMES
Chapter 843 - Obstructing Justice

843.16 Unlawful to install radio equipment using assigned frequency of state or law enforcement officers; definitions; exceptions; penalties—

(1) No person, firm, or corporation shall install in any motor vehicle or business establishment, except an emergency vehicle or crime watch vehicle as herein defined or a place established by municipal, county, state, or federal authority for governmental purposes, any frequency modulation radio receiving equipment so adjusted or tuned as to receive messages or signals on frequencies assigned by the Federal Communications Commission to police or law enforcement officers of any city or county of the state or to the state or any of its agencies. Provided, nothing herein shall be construed to affect any radio station licensed by the Federal Communications System or to affect any recognized newspaper or news publication engaged in covering the news on a full-time basis or any alarm system contractor certified

pursuant to part II of chapter 489, operating a central monitoring system.

This section shall not apply to any holder of a valid Amateur Radio operator or station license issued by the Federal Communications Commission or to any recognized newspaper or news publication engaged in covering the news on a full-time basis or any alarm system contractor certified pursuant to part II of chapter 489, operating a central monitoring system.

Any person, firm, or corporation violating any of the provisions of this section shall be deemed guilty of a misdemeanor of the second degree, punishable as provided in s. 775.082 or s. 775.083.

Indiana Scanner Law

IC 35-44-3-12. Possession of police radios –
A person who knowingly or intentionally:
possesses a police radio;
(2) transmits over a frequency assigned for police emergency purposes; or
(3) possesses or uses a police radio:
(A) while committing a crime;
(B) to further the commission of a crime; or
(C) to avoid detection by a law enforcement agency; commits unlawful use of a police radio, a Class B misdemeanor.

Subsection (a)(1) and (a)(2) do not apply to:
a governmental entity;
(2) a regularly employed law enforcement officer;
(3) a common carrier of persons for hire whose vehicles are used in emergency service;
(4) a public service or utility company whose vehicles are used in emergency service;
(5) a person who has written permission from the chief executive officer of a law enforcement agency to possess a police radio;
(6) a person who holds an Amateur Radio license issued by the Federal Communications Commission if the person is not transmitting over a frequency assigned for police emergency purposes;
(7) a person who uses a police radio only in the person's dwelling or place of business;
(8) a person:
(A) who is regularly engaged in news gathering activities;
(B) who is employed by a newspaper qualified to receive legal advertisements under IC 5-3-1, a wire service, or a licensed commercial or public radio or television station; and
(C) whose name is furnished by his employer to the chief executive officer of a law enforcement agency in the county in which the employer's principal office is located;
(9) a person engaged in the business of manufacturing or selling police radios; or
(10) a person who possesses or uses a police radio during the normal course of the person's lawful business. (c) As used in this section, "police radio" means a radio that is capable of sending or receiving signals transmitted on frequencies assigned by the Federal Communications Commission for police emergency purposes and that:
can be installed, maintained, or operated in a vehicle; or
(2) can be operated while it is being carried by an individual.
The term does not include a radio designed for use only in a dwelling.
[IC 35-44-3-12, as added by Acts 1977, P.L.342, SEC.1; P.L.162-1994, SEC.1.]

Kentucky Scanner Law

Crimes and Punishments
Offenses Against the State and Public Justice

432.570 Possession or use of radio capable of sending or receiving police messages restricted; penalty; enforcement

(1) It shall be unlawful for any person except a member of a police department or police force or an official with written authorization from the head of a department which regularly maintains a police radio system authorized or licensed by the Federal Communications Commission, to have in his or her possession, or in an automobile or other vehicle, or to equip or install in or on any automobile or other vehicle, any mobile radio set or apparatus capable of either receiving or transmitting radio or other messages or signals within the wavelength or channel now or which may hereafter be allocated by the Federal Communications Commission, or its successor, for the purpose of police radios, or which may in any way intercept or interfere with the transmission of radio messages by any police or other peace officers and it shall be unlawful for any car, automobile, or other vehicle other than one publicly owned and entitled to an official license plate issued by the state issuing a license to a said car, to have, or be equipped with the sets or apparatus even though said car is owned by an officer. This section shall not apply to any automobile or vehicle owned or operated by a member of a sheriff's department authorized by the fiscal court to operate a radio communications system that is licensed by the Federal Communications Commission or other federal agency having the authority to license same. Nothing in this section shall preclude a probation and parole officer employed by the Department of Corrections from carrying on his person or in a private vehicle while conducting his official duties an authorized, state-issued portable radio apparatus capable of transmitting or receiving signals.

(2) Any person guilty of violating any of the provisions of this section shall be guilty of a misdemeanor, and, upon conviction, shall be punished by a fine of not less than fifty dollars ($50) and not exceeding five hundred dollars ($500), or imprisonment not exceeding twelve (12) months, or both so fined and imprisoned.

(3) It shall be the duty of any and all peace officers to seize and hold for evidence any and all equipment had or used in violation of the provisions of this section, and, upon conviction of the person having, equipping or using such equipment, it shall be the duty of the trial court to order such equipment or apparatus destroyed, forfeited, or escheated to the Commonwealth of Kentucky, and said property may be ordered destroyed, forfeited, or escheated as above provided without a conviction of the person charged with violating this section.

(4) Nothing contained in this section shall prohibit the possession of a radio by:

(a) An individual who is a retailer or wholesaler and in the ordinary course of his business offers such radios for sale or resale;

(b) A commercial or educational radio or television station, licensed by the Federal Communications Commission, at its place of business; or

(c) An individual who possesses such a radio, provided it is capable of receiving radio transmissions only and is not capable of sending or transmitting radio messages, at his place of residence; licensed commercial auto towing trucks; newspaper reporters and photographers; emergency management agency personnel authorized in writing by the director of division of emergency management (for state personnel) or chief executive of the city or county (for their respective personnel); a person holding a valid license issued by the Federal Communications Commission in the Amateur Radio service; peace officers authorized in writing by the head of their law enforcement agency, Commonwealth's attorneys and their assistants, county attorneys and their assistants, except that it shall be unlawful to use such radio to facilitate any criminal activity or to avoid apprehension by law enforcement officers. Violation of this section shall, in addition to any other penalty prescribed by law, result in a forfeiture to the local law enforcement agency of such radio.

Michigan Scanner Law

Michigan Penal Code Chapter 750

750.508 Equipping vehicle with radio able to receive signals on frequencies assigned for police purposes; permit required; exceptions; misdemeanor; penalty; radar detectors not restricted. [M.S.A. 28.776]

Sec. 508. (1) Any person who shall equip a vehicle with a radio receiving set that will receive signals sent on frequencies assigned by the federal communications commission of the United States of America for police purposes, or use the same in this state unless such vehicle is used or owned by a peace officer, or a bona fide Amateur Radio operator holding a technician class, general, advanced, or extra class amateur license issued by the federal communications commission, without first securing a permit so to do from the director of the department of state police upon application as he or she may prescribe, shall be guilty of a misdemeanor, punishable by imprisonment in the county jail for not more than 1 year, or by a fine of not more than $500.00, or by both fine and imprisonment in the discretion of the court.

(2) This section shall not be construed as restricting the use of radar detectors.

Minnesota Scanner Law

299C.37 Police communication equipment; use, sale.

Subdivision 1. Use regulated. (a) No person other than peace officers within the state, the members of the state patrol, and persons who hold an Amateur Radio license issued by the Federal Communications Commission, shall equip any motor vehicle with any radio equipment or combination of equipment, capable of receiving any radio signal, message, or information from any police emergency frequency, or install, use, or possess the equipment in a motor vehicle without permission from the superintendent of the bureau upon a form prescribed by the superintendent. An Amateur Radio license holder is not entitled to exercise the privilege granted by this paragraph if the license holder has been convicted in this state or elsewhere of a crime of violence, as defined in section 624.712, subdivision 5, unless ten years have elapsed since the person has been restored to civil rights or the sentence has expired, whichever occurs first, and during that time the person has not been convicted of any other crime of violence. For purposes of this section, "crime of violence" includes a crime in another state or jurisdiction that would have been a crime of violence if it had been committed in this state. Radio equipment installed, used, or possessed as permitted by this paragraph must be under the direct control of the license holder whenever it is used.

(b) Except as provided in paragraph (c) any person who is convicted of a violation of this subdivision shall, upon conviction for the first offense, be guilty of a misdemeanor, and for the second and subsequent offenses shall be guilty of a gross misdemeanor.

(c) An Amateur Radio license holder who exercises the privilege granted by paragraph (a) shall carry the Amateur Radio license in the motor vehicle at all times and shall present the license to a peace officer on request. A violation of this paragraph is a petty misdemeanor. A second or subsequent violation is a misdemeanor.

New York Scanner Law

Title III - Vehicle and Traffic Law Article 12 - Other Provisions
Section 397 - New York State Vehicle and Traffic Law

s 397. Equipping motor vehicles with radio receiving sets capable of receiving signals on the frequencies allocated for police use.

A person, not a police officer or peace officer, acting pursuant to his special duties, who equips a motor vehicle with a radio receiving set capable of receiving signals on the frequencies allocated for police use or knowingly uses a motor vehicle so equipped or who in any way knowingly interferes with the transmission of radio messages by the police without having first secured a permit to do so from the person authorized to issue such a permit by the local governing body or board of the city, town or village in which such person resides, or where such person resides outside of a city, or village in a county having a county police department by the board of supervisors of such county, is guilty of a misdemeanor, punishable by a fine not exceeding one thousand dollars, or imprisonment not exceeding six months, or both. Nothing in this section contained shall be construed to apply to any person who holds a valid Amateur Radio operator's license issued by the federal communications commission and who operates a duly licensed portable mobile transmitter and in connection therewith a receiver or receiving set on frequencies exclusively allocated by the federal communications commission to duly licensed radio amateurs.

If you live in a state that restricts the rights and freedoms of its citizens, or if you have a scanner in your vehicle and plan to travel through one of these states, it is important to be aware of these restrictive laws.

It is interesting to note, however, that in most cases a person possessing an Amateur Radio operator's license is exempt from the provisions of these antiscanner laws. This is just one more advantage of getting your Amateur Radio operator's license.

In addition enabling one to monitor the police and fire frequencies, a scanner makes it possible to listen for traffic on any number of other frequencies. I keep one bank of frequencies on my scanner programmed for the Amateur Radio national calling frequencies. Another bank of frequencies contains all of the FRS/GMRS frequencies, while a third bank contains all the CB radio frequencies. My scanner already has the NOAA weather frequencies built in, but if yours does not, this is another useful bank of frequencies to program so that you can listen to the NOAA forecasts and alerts.

When programming a scanner, there is a significant advantage to programming groups of frequencies into separate banks. Because a bank of frequencies may be selected as a priority group of frequencies, or locked out when not wanted, one can listen to those channels and frequencies that are most important or most interesting at any given time.

Field and mobile communications are the heart of communications for survival and self-reliance. The ability to set up worldwide communications with no more equipment than can be carried in a briefcase or daypack means that when the communications infrastructure fails or when we travel to areas where that infrastructure does not yet exist, we can still have reliable communications.

With our mobile receivers and scanners we are able to monitor public service and emergency units to obtain firsthand information about conditions in our area. When we can communicate from the field or from our vehicles as we are mobile along some highway or back road, we can communicate from anywhere.

CHAPTER 7

Power Supplies

Your radio equipment will either be powered from the AC power lines of your home or run off batteries. Many "base station" radios come with a built-in AC power supply. Simply plug the cord into the wall and you have power for your radio—assuming that you have power available from the outlet to begin with. Other radios—some "base stations" and all "mobile stations"—require a 12-volt DC current to operate. The radios requiring a 12-volt DC power supply can be connected directly to a 12-volt battery. When operating these 12-volt radios from home, it is usually more convenient to connect them to a 12-volt power supply that itself draws power from the AC lines coming into your home. Of course, you can also simply connect the radio to a 12-volt battery sitting in your living room.

Various power supplies and a solar panel.

For survival and self-reliance communications, all radios should be capable of operating off batteries or other noncommercial power-line supplies. Your survival communications become useless if they are dependent on commercial power and you are in the middle of an extended power outage.

SOLAR POWER

Solar power is a useful method for maintaining a constant working charge on your battery packs and for running some of your equipment. The effectiveness of your solar power system depends on the size and efficiency of the solar panel itself and on the amount of sunlight available in the area in which you are located.

Using a small solar panel, I have found that it is a fairly simple matter to run a radio in receive mode. Using that same solar panel to run a radio transmitter is more problematic. Simply put, it takes much more power to transmit than it does to receive, and small solar panels just don't have the power output capability to directly power a transmitter.

Small solar panels are very useful for maintaining a constant trickle charge on a battery pack. By keeping a solar panel always connected to your battery pack, there is a constant flow of power to the batteries as long as the solar panel is in direct sunlight. Depending on the efficiency of your solar panel, it could take several hours to replace the power you used during an hour of on-the-air communication. However, if you use your radio to make a contact every evening and let the solar panel recharge the batteries throughout the day, you will be able to maintain this procedure as long as the sun continues to come up each morning.

A 10-watt solar panel.

OTHER POWER METHODS

Solar panels are excellent tools for powering your radio equipment and charging batteries when you have an area with good sunlight. However, at night or under poor weather conditions you still might want to charge your batteries and run your communications equipment.

An excellent tool for charging batteries is the "Motorola FreeCharge" hand-crank charger. The FreeCharge charger is designed to charge cellular telephone batteries. Currently the FreeCharge can be ordered with connectors for the various

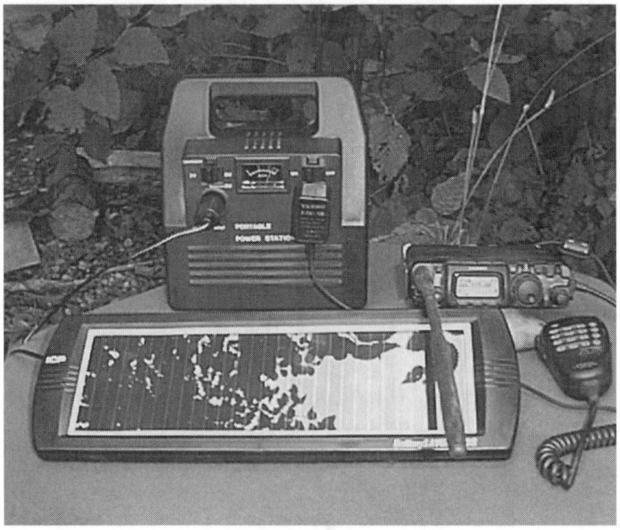

Mini power station—solar panel and Yaesu FT-817 radio.

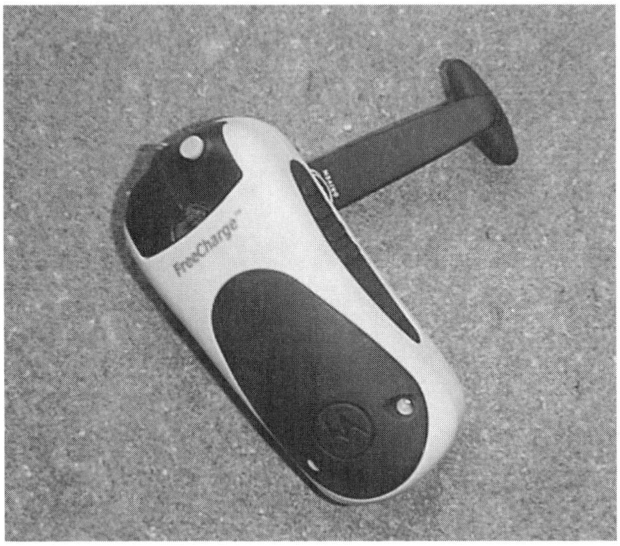

Motorola telephones and for Nokia telephones. It is a simple matter to adapt these connectors to work on your radio equipment—simply wire the appropriate connector to fit the charging port of your radio.

Another advantage of the FreeCharge is that it stores power in its own internal power cell, allowing you to use the FreeCharge as an external battery pack. When not using the FreeCharge to power your communications equipment, you can plug in the FreeCharge's flashlight module and use it to light your way.

The Motorola FreeCharge functions similarly to the various self-powered radios on the market, except that the FreeCharge can be used to power most any type of cellular telephone or radio for which you can wire an adapter.

BATTERIES

Batteries will be your primary power source for survival communications and will serve as a backup or alternate means of power for your communications during other times, thus making your station self-reliant.

First we will look at batteries intended to be used internal to the equipment they are powering (e.g., batteries in a flashlight).

Alkaline batteries are the popular nonrechargeable batteries most commonly available and used today. Nickel cadmium batteries are the older version of rechargeable batteries. Nickel metal hydride batteries are a newer version of rechargeable battery. Lithium ion batteries are the newest version of rechargeable battery. These batteries provide a good long-lasting charge, very suitable for running radio equipment. Lithium ion batteries cannot be overcharged and do not lose their charges when stored.

Vehicle batteries are intended to provide a high current for a short period of time in order to start a vehicle (or other like function). Once the battery has provided the short burst of energy needed to start the engine it is not thereafter needed to keep the engine running. The running engine provides the power to recharge the short burst of energy used from the battery. This short burst of energy does not, however, deeply discharge the battery.

While a common vehicle battery can be used to power your radio equipment, it is not ideally suited to doing so if it is not receiving a constant charge from your running automobile engine. Deeply draining and recharging common vehicle batteries will greatly shorten the useable life of the battery in question.

Deep cycle batteries are similar in appearance to vehicle batteries but are specifically designed to accept deep discharge and recharge (deep cycle). This type of battery is commonly used to power equipment such as golf carts and trolling

Solar battery charger.

motors. Deep cycle batteries are a good choice for providing external battery power for your communications equipment.

A radio without power is just a very expensive paperweight. To successfully communicate in survival and self-reliance situations, you must be sure of having sufficient electricity to run your radios. We have seen that batteries will be the primary source of power when commercial power fails or when you are operating in remote areas. We have also looked at ways to keep those batteries charged and ways to provide power directly to your radios from systems such as solar.

When making your communications plan, remember the power source.

Rear view of solar battery charger, shown charging four AA batteries.

CHAPTER 8

Antennas

Although it is your radio that produces a signal, it is your antenna that sends that signal to others, and it is your antenna that captures signals from the airwaves and sends them to your radio where you can receive them. All radios, whether transmitting or receiving, need some sort of antenna in order to function.

Most radios use a single antenna for both transmitting and receiving. This is especially true for single-channel operation, referred to as simplex or one-way-reversible operation. However, a radio can also be set up to use one antenna for transmitting and another for receiving. This is common when operating a separate transmitter and receiver or when operating duplex—that is, sending on one frequency and receiving on a different one.

The specific function of the antenna depends on whether it is transmitting or receiving. A transmitting antenna converts the radio frequency output of the transmitter into an electromagnetic field that is radiated into space. A receiving antenna captures this electromagnetic field and converts it into a radio frequency signal that is delivered to the receiver. The gain of an antenna is primarily a function of its design. Gain is a calculation based on improvement over a quarter-wave vertical antenna. A quarter-wave vertical antenna is considered to have zero gain. When you purchase or build an antenna that is said to have gain, that gain is described in decibels (dB).

Although you will still be transmitting with the same number of watts, your effective radiated power (ERP) is increased by the gain of the antenna. For example, if you are transmitting with 2 watts of power, and you connect your radio to an antenna with a 4 dB gain, your ERP is 5 watts (2 x 2.5 = 5). The use of the gain antenna gives you equivalent radiated power with 2 watts as if you were transmitting with 5 watts into a quarter-wave vertical antenna.

The following chart shows the multipliers for antenna gain in decibels, allowing you to calculate your ERP based on your antenna and the power output of your radio:

Gain in dB	Multiplier
1	1.2
2	1.6
3	2.1
4	2.5
5	3.0
6	4.0

7	5.1
8	6.3
9	8.0
10	10.2
11	12.6
12	15.9
13	20.0
14	25.1
15	31.6
16	40.0
17	50.2
18	63.3
19	80.0
20	100.0

Radio signals radiated from an antenna have a specific pattern. A vertical antenna radiates signals equally in all directions. A horizontal antenna (such as a half-wave dipole) is basically a bidirectional antenna, while an antenna such as a half-rhombic is unidirectional, radiating the bulk of its energy in one direction.

It is important to understand the radiation pattern of the antennas that you will be using. If you will be communicating with mobile stations or want to be able to communicate with all stations in a general area, you will want to use an antenna that radiates its signal in all directions and receives signals equally well from all directions. If, on the other hand, you want to communicate with a single fixed station, a directional antenna might be your best option. It sends the bulk of your signal directly toward the intended receiving station and reduces interference from stations located to the side and rear of the transmission path.

When building antennas, it can generally be said that bigger is better than smaller, and higher is better than lower. With an antenna intended only for receiving, it might be sufficient simply to run a long wire from the peak of your roof to a tree across your yard. This will certainly provide better reception than the telescoping antenna on the back of your radio.

For an antenna intended for transmitting, it is necessary to match the antenna to the frequency being used. While antenna theory can become fairly complex, for our purposes we will keep things simple. We will discuss building half-wavelength antennas. To calculate the length of wire required for a half-wavelength antenna, divide your intended operating frequency into a constant of 468. (The number 468 relates to the speed of a radio wave through copper wire, which is about 10-percent slower than the speed of light.) So if you want to build an antenna to transmit on the 20-meter Amateur Radio band (14.000 MHz – 14.350 MHz), choose a frequency in this band for your calculations. This frequency is either chosen at the center of the band to give the best characteristics across the entire band, or if you have a specific operating frequency, use that. For this example we will choose 14.225 MHz. Thus, we have 468/14.225 = 32.89982 feet, or, for simplicity's sake, 33 feet.

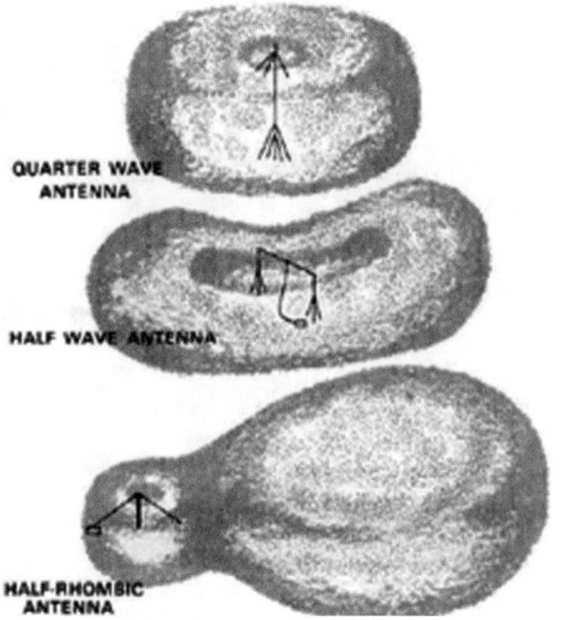

Examples of antenna radiation patterns.

CONSTRUCTING A WIRE ANTENNA

Constructing a wire antenna is a fairly simple matter. Begin with two lengths of copper-wrapped steel wire, three insulators, a length of

450-ohm transmission line, and, if desired, an adapter to connect the transmission line to a coax connection.

In this case, we are using 66 feet of antenna wire cut at its center, making each side of our antenna 33 feet long. The transmission line is 17 feet long. Using an antenna tuner, we will be able to tune this antenna to resonance on all bands from 10 to 40 meters. Impedance at the end of the 450-ohm transmission line will be 50 to 60 ohms, making it suitable for radio transmission and reception.

To construct the antenna, first connect an insulator to one end of each of the pieces of copper-wrapped steel antenna wire. Connect the opposite ends of each of these wires to the center insulator. Each piece of antenna wire should be separated from the other by the center insulator (i.e., one half of the antenna does NOT touch the other half at the center connection).

Now connect one side of the transmission line to each half of the antenna at the center insulator. This connects each side of the antenna through the transmission line. This connection is best made by wrapping the antenna wire and transmission line wire around each other and soldering the connection.

At the opposite end of the transmission line, connect a coax adapter, which will allow you to run a length of coax cable from your radio to the antenna.

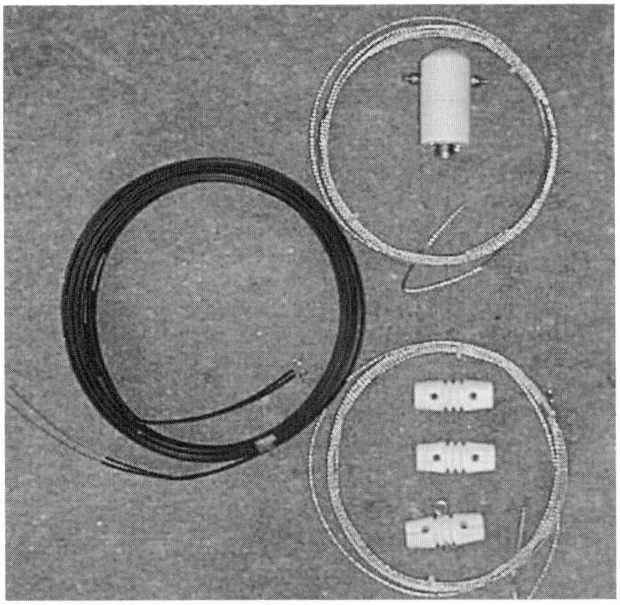

All you need to construct a highly effective wire antenna.

Attach a rope to the each of the insulators on the ends of the antenna and stretch the wire between two supports (e.g., trees, buildings). You should try to get your antenna at least 20 feet off the ground (remember, higher is better).

This half-wave horizontal antenna will radiate broadside to its length. So . . . you should hang your antenna to project this radiation pattern in the direction best suited to your communication needs. Two of these antennas can be used to cover all directions by hanging one running north-south and one running east-west.

While your antenna can be hung horizontally, it can also be used in other patterns. You can use your antenna as an inverted "V." Hang the antenna by its center from a single support, such as a tree. Angle the ends of the antenna wire down and out, forming an inverted "V." The inverted "V" antenna is a directional antenna radiating its energy in the direction of the downward slope of the antenna. Slope the open end of the "V" in the direction toward which you wish to send and receive signals.

You can also hang your wire antenna vertically and thus have an antenna with a 360-degree radiation pattern. Simply attach a long guy wire or rope to one end of your antenna and pull it up onto a support higher than the length of your antenna. Attach a short rope to the bottom of the antenna and secure it to a stake in the ground directly below the upper support for your antenna. Your radio is still fed from the center of the now vertical antenna. While this vertical arrangement does work well as an omnidirectional antenna, its downside is that it requires a fairly high support, especially when working on the lower frequencies. For example, if we were to use this same antenna in the 40-meter band, we would require a vertical support more than 65 feet high to use it as a vertical antenna. However, if we were to use this antenna in the 10-meter band, we are looking at an overall length of only about 16 feet.

Wire antennas used in their various configurations are easily constructed and greatly

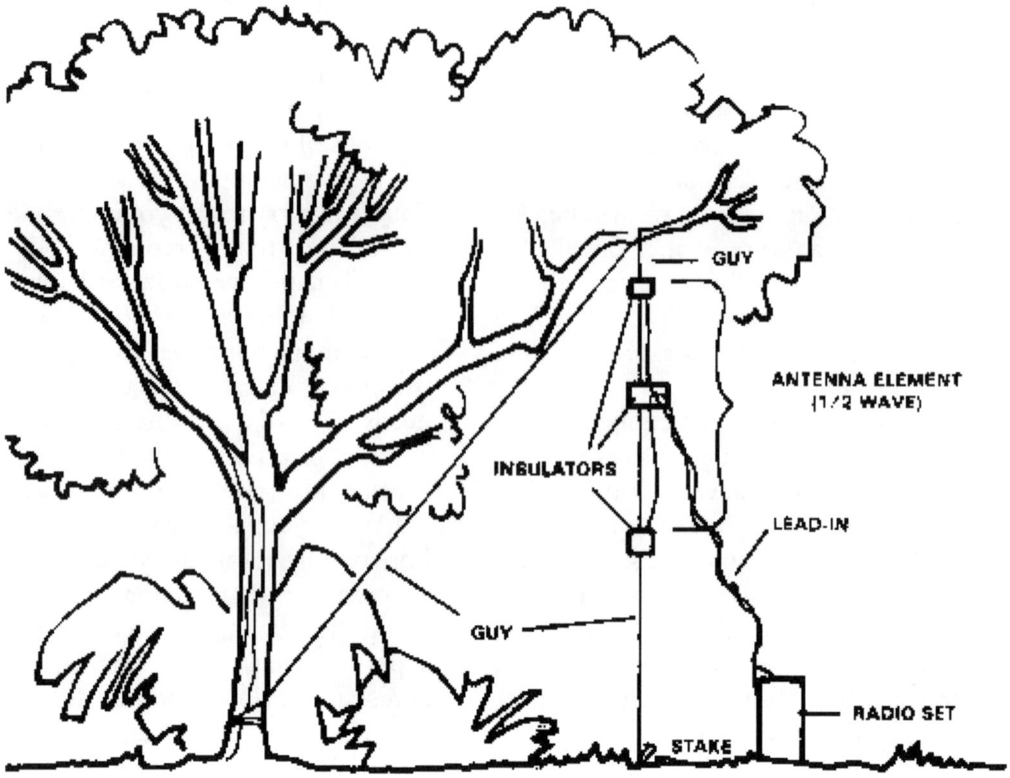

One method of installing a wire antenna in the field.

improve your radio transmission and reception. Another advantage of the wire antenna is that it's unobtrusive. A wire running between two trees in your backyard or sloping down from the peak of your roof attracts little attention. The antenna can also be carried easily in a rucksack or briefcase and put up at a campsite or other temporary location.

ARROW ANTENNAS

In addition to operating on HF bands, you will also want to have an antenna to increase the range of your UHF/VHF communications. The short "rubber duck" -type antennas that come as a standard on most handheld/walkie-talkie radios are a compromise between size and efficiency. Because these radios are designed to be carried in a pocket or clipped to one's belt, large antennas simply are not an option for regular operation. However, there are certainly times when you will want to increase the communications range of your walkie-talkie or other small UHF/VHF radio while maintaining your mobility.

The absolute best option I have found for increasing the efficiency of UHF/VHF communications from portable radios is the inclusion of an Arrow Antenna (www.arrowantennas.com).

The Arrow Antenna is a handheld antenna specifically designed for field operations. It provides a gain to your signal output and is directional, thereby letting you focus your signal toward the area with which you are trying to communicate.

Arrow Antennas are constructed of a lightweight aluminum boom and aluminum arrow shafts for elements of the boom itself. The whole antenna can be disassembled and packed in a roll-up bag for easy transport. This allows you to have your Arrow Antenna immediately available.

You can operate your radio with the Arrow Antenna attached in a portable mode (walking and talking), holding your radio in one hand and the antenna in the other. More efficiently,

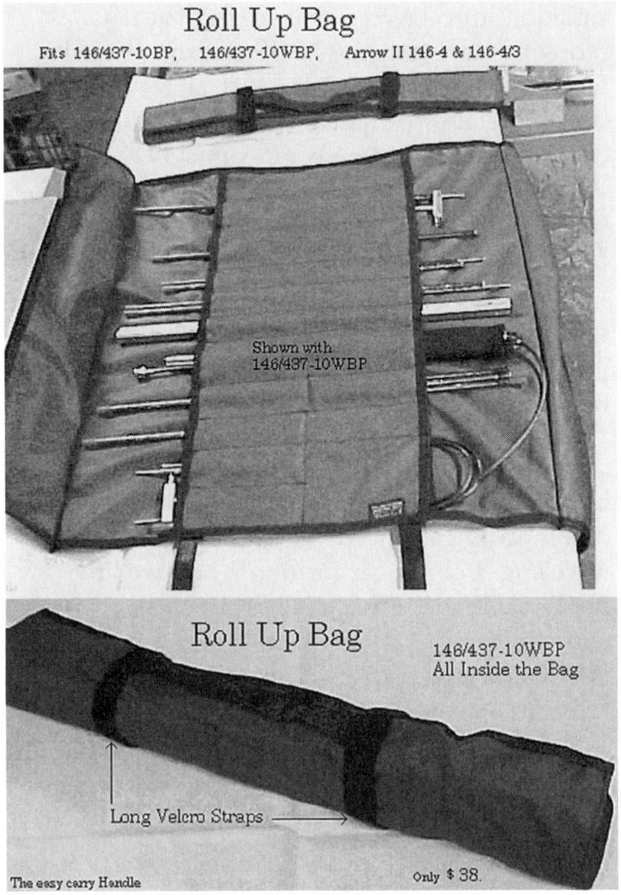

The amazing Arrow Antenna. (Photo courtesy of Arrow Antennas.)

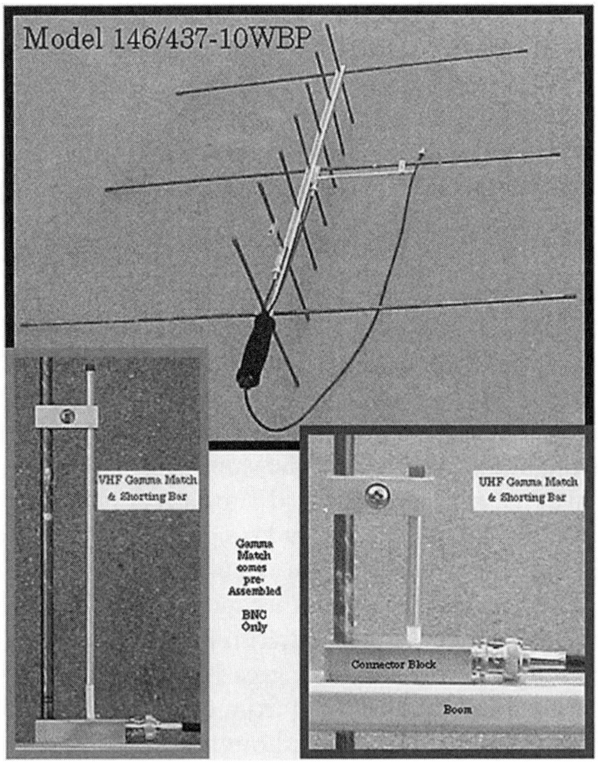

The Arrow Antenna in its bag ready for transport. (Photo courtesy of Arrow Antennas.)

however, you would use your Arrow Antenna during stops or when setting up a camp and talking back to a staging area or back to home. On several occasions while operating radios from the backwoods, I have found it impossible to communicate on UHF/VHF frequencies using just a "rubber duck" antenna, but after connecting my Arrow Antenna have had loud and clear communications with several distant stations. I often simply just point my Arrow Antenna in the direction I want to communicate and do just fine. On occasions when I am camping in an area for a couple of days, I attach the Arrow Antenna to a tree or branch, pointed toward the area with which I wish to communicate.

If you intend to use UHF/VHF radios for communication from remote areas, it is important to remember that your "rubber duck" antenna will only provide communications over a limited range. To increase your communications range it is necessary to use a better antenna.

NVIS

The ground wave portion of an HF radio wave becomes unusable at a distance of about 50 miles—and often at much less distance, depending on frequency, antenna type and polarization, and power output of the radio. The sky wave portion of the same radio wave, using standard antennas, is not reflected back to earth at a distance less than 100 miles. This can leave a gap of 50-plus miles where HF radio communication is ineffective (the skip zone).

NVIS (Near Vertical Incidence Skywave) is a communication technique for high-frequency communications using antennas with a very high angle of radiation, often up to 90 degrees (straight up). When combined with a frequency below critical frequency that will be reflected off the ionosphere (generally in the 40-meter, 75/80-meter, and 160-meter bands), it provides for highly reliable communications in a roughly 200-mile circle around one's position – with no skip zone. Depending on takeoff angle and critical frequency, this circle can be greatly expanded.

Using the NVIS technique, the radio signals are refracted off the ionosphere and return to earth in an omnidirectional pattern at all angles, resulting in no dead spots in the reception pattern of the signal. The effect can be compared to spraying a water hose with a fogging nozzle straight up into the air. The water falls back to earth in an evenly dispersed circular pattern out to a given distance. The radio signals from NVIS return to earth in a similar circular pattern.

Because NVIS signals are radiated in a near vertical pattern from the transmitter the effects of terrain and vegetation are greatly reduced. NVIS signals can be transmitted from deep valleys or the bottom of canyons without any major variation in received signal strength at any receiving station in the NVIS returned signal footprint. Furthermore, because all returned signals are approximately the same strength, radio direction-finding of an NVIS signal becomes very difficult. Time of arrival is almost exactly the same, and the *actual* direction toward the source of the incoming wave is "in the sky." This is a *very* effective anti-DF (direction finding) technique, particularly when transmitting from a deep canyon that produces virtually NO usable ground wave to DF.

Antennas that work well for NVIS communication are low dipoles, full-wave horizontal loops, and inverted "V" antennas. A dipole strung at one-tenth to one-quarter wavelength above the ground works well for NVIS. Antennas hung even lower (between two and ten feet above the ground) also work well for NVIS, although the lower hung antennas may result in weaker signals. An inverted "V" with an apex angle of 120 degrees or more also works

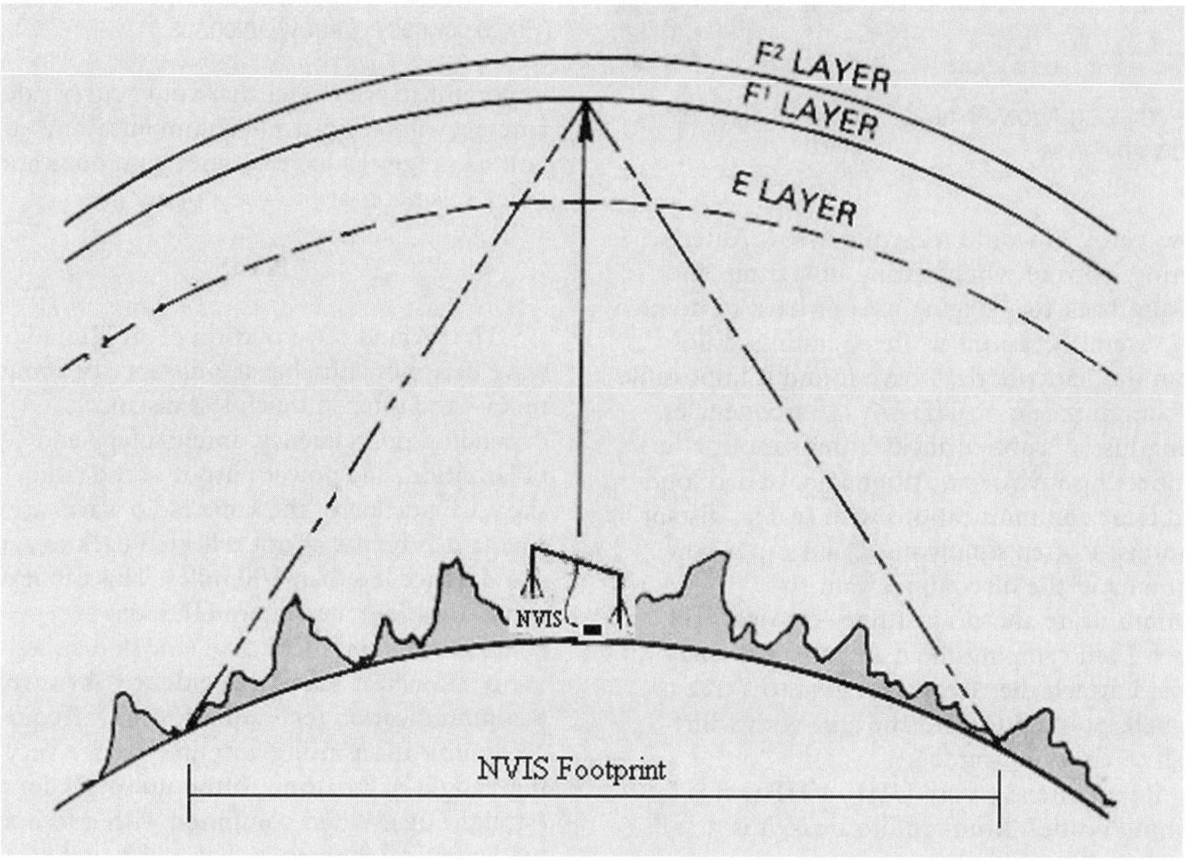

very well for NVIS and can be erected using only a single support at its center.

When we consider horizontal antennas (i.e., dipoles), we generally make sure that the antenna is broadside to our intended communications path. However, with NVIS, orientation of the antenna does not matter since the signal is radiated up with the intent of refracting it off of the ionosphere. The NVIS antenna may be strung wherever the terrain makes it most convenient.

Selection of an appropriate frequency is essential when using NVIS. In all sky wave propagation there is a frequency above which the signal will pass through the ionosphere and on into space and will not be reflected back to earth. Obviously, if the frequency is not reflected back to earth, it is of little value for NVIS radio communications. However, this usually isn't a major problem. A selection of three frequencies will usually permit ongoing NVIS communication. In the low end of the solar cycle, NVIS frequencies are going to be limited to the 160- to 40-meter bands. At other times, NVIS communication *might* be possible on the 30- and 20-meter bands too; however, this is not guaranteed. Therefore, we selected our NVIS operating frequencies from the 160- to 40-meter bands.

Generally, three NVIS frequencies will be sufficient. You will need a daytime operating frequency, a night-time operating frequency, and a transitional frequency to use around sunset and sunrise. I use a 40-meter frequency during the day, a 160-meter frequency at night, and an 80-meter frequency during transition. I have also found that the 80-meter frequency is often useable throughout the night.

NVIS isn't the answer to every communications problem, but it is a very effective technique for maintaining regional communications. NVIS antennas are easy to set up, can be packed easily, and as long as you have chosen a proper operating frequency, are very efficient.

If you carry HF radios, NVIS is a technique that should certainly be added to your survival and self-reliance communications package.

A good antenna system is the key to effective radio communication. The ability to hear a distant or weak transmission is dependent on the antenna. The ability to transmit a signal that can be received by others is dependent on the antenna. We have looked at various antenna systems and their uses, but if there is one thing to be learned from this chapter, it is to use the best possible antenna available when communicating by radio.

Antenna structure and function could fill a lifetime of study, and while you may not wish to devote a lifetime to this pursuit, it is essential that you understand the basic concepts.

CHAPTER 9

Codes and Ciphers

Codes and ciphers, whether they are performed with a pencil and paper or with the most advanced computer, all serve the purpose of protecting our privacy and safeguarding our secrets. When establishing a communications network for survival and self-reliance, you will often be relying on radios for communication. Because anyone can monitor radio communications, you might want to consider the use of codes and ciphers to safeguard your communications.

In some circumstances (such as Amateur Radio), the use of encryption is proscribed by government regulations. You certainly want to follow applicable regulations and procedures to keep a sense of order and structure to crowded bands and frequencies available for use.

When considering the use of codes and ciphers, remember: people are only truly free to speak their minds in open debate when they are also free to speak privately without fear of government surveillance and censorship.

ENCRYPTION PROGRAMS

With the advent of the personal computer, we are provided with capabilities in our homes that no more than several years ago were available to only the largest corporations and government agencies. One of those capabilities, and the one we will focus on regarding communications for survival and self-reliance, is encryption.

Using our home computers and freely available programs, we have the ability to perform high-level encryption that is unbreakable. Now, it must be understood that while there is only one type of encryption that can be mathematically proven to be unbreakable (the one-time pad), some computer-based encryption is, for all practical purposes, unbreakable using current technology.

The advantage of computer-based encryption is that it is extremely fast when compared with any type of pen and paper cipher. Furthermore, computer-based encryption can be incorporated into our regular communications, thereby adding an immediate increase to our overall communications security with no more effort than a keystroke or a click of your computer's mouse.

E-mail is regularly monitored as it passes across the Internet. Your employer could be monitoring your e-mail at work, and the courts have ruled that he has a right to do so. In fact, the federal government, with its programs such as Carnivore (DCS1000) and Magic Lantern, could very well be reading your e-mail and other electronic communications.

A very important part of survival communications is the ability to send messages that are only capable of being read by their intended recipients. In my other books, *The Complete Guide to E-Security* and *Freeware Encryption and Security Programs*, I discuss computer-based encryption in detail and explain how to use it to secure your communications on the Internet.

The Internet is a valuable tool for survival and self-reliance communications, because it has built-in multiple redundancies, making it very difficult, if not impossible, to shut it down over a wide area or for any extended length of time. Furthermore, the type of communication that can be sent over the Internet (e-mail, file transfer, etc.) can be sent via other means, such as packet bulletin board systems (PBBS) or file transfer using such protocols as G-TOR.

Because of the ability to use computer-based encryption to secure messages and files that can be sent by multiple means, we will look briefly at two computer-based encryption systems (PGP and ABI-Coder) and a Web-based secure e-mail system (ZipLip).

Pretty Good Privacy

Pretty Good Privacy (PGP) is a high-level encryption program developed by Phil Zimermann and released to the public in 1991. Since that time, PGP has undergone regular upgrades and improvements, and it has become the unofficial standard for communications security on the Internet.

If you have a computer, you should have PGP installed on it! If you send e-mail, it should be encrypted using PGP (or another high-level encryption program). PGP is available as freeware from many places on the Internet. The two most common places to obtain your free copy of PGP are (1) the Massachusetts Institute of Technology PGP download Web site at http://web.mit.edu/network/pgp.html and the PGP International Web site at www.pgpi.org.

The following is an example of PGP encrypted text:

——-BEGIN PGP MESSAGE——-
Version: PGP 6.0.2

qANQR1DDDQQDAwL8FuAWCXYQ2mDJwTftxMm58OTyrx0/liSCHOgBpkY26/QIGN3V
7/sdnUuL3bXTwHPM73QgvcMbwRy+1ndEJzfPBzGK9ZbxrjgHB2Aty9ToLOuqy+0V
lIy+O8oU1yP+uhE5k3F7WAkKadOtCDzFer3eGg6KN6W+tW8V2R8085dxY8EqnRWW
G4+0wN+rOb/tYfbLhVpbYltWPtYrBFAl9ziuEvnuurH/j1m+Vc5HnQAZ+gw0kaR/
eUSWWZumYVj5i8GgglGwZshmPMzxxPAmAwvoOlV77GEETHGvHY4kIYr4zT+i7LWe
nRwVt0xe8Z6ut9g+Dr4Dcbeo/dEsnc0fCmgtU8bmB0p3xuTlnYxX2lDAr8yIWbTG
crkodFpvi/pJmm1ufdHHDuqHhD4t4VmmwKL8o2vxgZ9jz4ZfiSc2tIwP3cLMiyhI
qRCqhRIBo2zGsAoA9TScVRsZFagIWmXvu197EuOqHuWVhOavJDzkKhi4WtuRqwSm
hYdHDJpzP+6Nfjifhyhv4plFW5Mgq59FeBznt0B/6HbDeYKygLiPxNmx8lvxXcFM
wSM2kJaYYbHsMymzXpK71MmzpfheWwXLs4LBtC7llxlYumr1lkqIvQRL4JwwjNp4
Jby1mNeNOAp8Ku7Z3iLoVWxz2LqwhEdoMQMA67G77CqE1W1xH4w73Sd/a59g9A==
=CQWq
——-END PGP MESSAGE——-

ABI Coder

ABI Coder is a high-level symmetric encryption program that I recommend you have available for your communication security needs. ABI-Coder is available as freeware from the ABI Web site at www.abisoft.net/bd.html. ABI-Coder uses the 448-bit BlowFish algorithm and the 168-bit 3-DES algorithm for encryption, making it a very strong encryption program.

One of the major advantages of ABI-Coder is that it allows the creation of self-decrypting files. You can encrypt a file and send it to someone who does not have ABI-Coder installed, and that person will still be able to decrypt the file and read its contents, providing that he knows the password used to secure the file when it was encrypted.

The following is an example of ABI-Coder encrypted text:

ABICODER“ª@¼Ï˜Ãæ□
U'Ô‰øÁ□□4WùŽ±á?B□□Ï'mµ¡□‡3z·8`’¿®+[Ë□+2æ½GyxÓÔ¾Â¶+□î»ÅÕ+ð®□□'²…ÅÌv
,□&□½ §`□UÜ□□$Ìó□õw□9q©□□¾!ÞÙ…s÷FŸb□-ûM')OþÆ□o¸NžC¯Ó□IA³,Á
·->(Å□$B!ØT_]tF'šl8æŽ(ñ□□}¨7‹□qaP½Z4?jíLó»□øúm™”□Õj□³Ûè¨åôàÑIæl37ž¯•□Ð³R"ÂÎ
□Fs#ç□“Ùë”ãÖA□□[½¢:ÕYB□Ø¿¢¡-ËðëÃÉÐ·àþíô[â/çIÌ /UC–ø‹oyY=ù}¹
<YóKÖfÿÆÕc*~°U?+™˜□^~□,CTçHèË□ˆøÑ;lÂà-² Ó)ÜyEUQÍöè□SÉ□)âA□×*·—
`□÷×0¾’□□-|□IÁgj©y¹□¦i□Õ‰¡£;M□ë0□¨□Bß¬ÈôÛ%àñŒ□xÛÊP^¹Ç]Öµ@b“Bw>Xšçí?Ëþ‹t’¿®+[Ë
□+2æ½GyxÓÔ¾Â¶+□î»ÅVIüVÈ™□m›zv\ésà°É€□½zd>ÌÏ$]8/€/IþÚh›ïÈû)|¡7 ã§-í´7-
) :È3%†ž¢□•Õ„±□o[·np£³AyÊ¦´›&W□¯□}Ô-Ë‘□ô²7Úé¶”>8¤'œÄtâ„aé>Öç□¥ÀVÎÉû‰%B'×à¸¯ì_úM½□
□yïnNíõ²□e8*7ZÞ□« Tô-Q*ÔÄ□¢ü□ò□E™y¤æ0#□□ˆ?gjütÓ§ùG·`è(áñ>§ôxÀaŸV□Dã+¹èÕ□$<%ä□
Ç}‡□YM“±’LQi½QÕÒ¨□q»ýä«l□□–“¯?šk½)…åe5¸ŒÈß› ì´@i¦Dð9ö÷Z£□×Çf|□Gâd…□□
|ÞT®S\;ÇA;ó/^□Ó□oÙA4”Â®³½□-ðÕ□□FwD-¤Ž¦„!p9Ò®?ÇU7.¥¹⁄26^ë– kÈw□€6¦2‘îŠd□/1

Super Encryption

If you have a file that simply must remain secured against all known attacks, you can use a technique known as “super encryption.” Super encryption is a procedure where you encrypt your message using any standard encryption method. You then take the output of your encryption (the ciphertext) and encrypt it again using a new key, or maybe even a completely new encryption method all together. The likelihood of anyone’s being able to crack a high-level encryption algorithm—such as those used by PGP and ABI-Coder—is extremely low. However, if you foresee a future weakness in any algorithm, simply use super encryption to protect your most sensitive information.

National-level government agencies are the only ones with any potential to defeat a program like PGP or ABI-Coder. However, if this is a concern, super encryption will stop even these agencies from decrypting your messages.

ZipLip Free Secure E-mail

Many people use free e-mail accounts (e.g., Hotmail or Yahoo), even if they have another paid e-mail account. There are several good reasons for using these free Web-based e-mail accounts, such as preventing criminal spammers from disrupting your paid account, having access to e-mail while traveling, and having an e-mail account offering some degree of privacy over your paid account.

If you are going to use Web-based e-mail, there is one such service that I strongly recommend. That service is ZipLip at www.ziplip.com. ZipLip has several security enhancing functions, but the feature that makes it unique is the way that it handles your secure e-mail.

When you send a secure e-mail using ZipLip, you compose and address your e-mail as you would any other type of e-mail. You then assign a password to that e-mail, and you have the option of including a password hint with the message.

When you press "send," your message is encrypted and stored on the ZipLip servers. The addressee of your message is sent a message containing a link to that message on the ZipLip servers. When the recipient of this message clicks on the link to your message, he is asked to enter a password to access the message. If he enters that correct password, he is taken to your message where he can read it, copy it, and so on. If, however, he does not enter the correct password, he is denied access to that message. Once a message has been successfully accessed on the ZipLip servers, it is deleted within 24 hours.

Password Hints

One problem faced when using symmetric (single-key) encryption programs is letting the recipient of your encrypted message know what key (password) is needed to decrypt and read the message. Many encryption programs (e.g., Ziplip) provide a place for you to include a password hint. Providing a hint to a password significantly reduces the security provided by the encryption program securing your message.

However, unless you have arranged a password list in advance, you will need to provide the recipient of your message with some clue about the password securing your message.

A method I have used to provide password hints is a table containing random sequences of numbers and letters. The table must, of course, be provided in advance of communication, but because of the large number of possible password combinations that can be created using this type of table, it has fairly long-term usability.

The advantage of using the random sequence table is that it allows construction of a large number of passwords from a single sheet of paper, as opposed to having multiple pages of predesignated passwords.

	A N	B O	C P	D Q	E R	F S	G T	H U	I V	J W	K X	L Y	M Z
A -	zgS2	1KPX	DJDn	i150	866t	jFRy	1IsK	NIOU	Qt60	T4WY	MPCw	Ckcc	ftjC
B -	iDUN	vhDx	4IRQ	6glj	THrg	gsFG	x6ps	qvj6	3BKE	yUXM	0zuS	dfwz	emnP
C -	q9f3	x1X5	0XRh	n5su	B8Jz	d9XT	Qvfw	x8u6	9ytw	1psp	75Dv	cCNw	KUSi
D -	e0Hx	pA7E	nqrd	KiZ1	J2md	9CE1	xL0i	vETu	betd	q3Lh	G1R7	gfz7	bDXk
E -	1HEW	SX30	MIz1	2afa	kt0s	miu3	PsZT	z0hL	AEqc	jg05	V1WE	fbUO	GWXi
F -	qUZX	B1pR	RuHo	KZuL	CgCH	KxMS	tw4g	gMui	5tcs	ybR1	DP3S	LZEA	KgpY
G -	sgsT	XVRt	avA5	hHDN	FQdh	iZIH	9oQy	5dXT	Xu2U	ubnA	SMgL	pAkP	CGjm
H -	geZC	Ly9p	a4pm	TT2o	xkQh	pLm5	W8Xh	Mq6Z	R7zI	PmVq	3AEo	NFP1	rj1B
I -	TEEw	10Yu	YzCj	gxUY	6LgJ	10DD	NP9x	izDa	2T8x	jGYN	p7yR	B1Xd	7bw6
J -	j9Nr	K0Gk	aFK4	kBv2	PiOH	vuPt	4p24	x3jL	sNz1	FqVt	iLKI	R3ik	OPcp
K -	oETB	6WQX	9s8I	TnZ1	1RIb	1JCw	vWjJ	yoHv	YOBO	L84Z	c5YU	63xi	QU4f
L -	vkPN	O6Gx	OAzs	QZxu	01wP	L6cE	MxQv	GIey	GRdf	K0ZN	SFKU	9SF2	bZT2
M -	GTM2	UBC4	vh4A	YAqs	R8JU	LCJS	gGNm	JGkf	Rtdj	KO6s	35Tg	YsYT	03xW
N -	ejMi	qtf1	dnnI	UTnW	ZTan	UkSt	vwJB	LWiG	iXDS	bPmm	DfNX	AEbN	56qz
O -	ZORb	KqSr	h2TY	AyP3	GhOn	znjp	OsRk	jUMP	Z5U1	bzDp	81B8	nvlS	CPxS
P -	Zbok	HnfN	DUVJ	LAGP	JpsJ	vdKY	U1Fd	YCrm	s9eb	cKNU	Gtbu	PgEt	QCKC
Q -	5kna	kz4W	UAon	0fyO	XFGD	9p4J	7wEK	Ptu4	20rd	1Rwx	QUQA	KH8n	LYmE
R -	Jksg	6G6L	RaGN	o0YZ	gcl1	cS3v	vyFf	mSbb	JKEh	1Aly	tsmz	A98j	1iLd
S -	8axg	AnYa	LfJ0	zobB	WZOw	5f0s	G2vD	aBZK	m9bt	aO4i	zrF6	yZq8	UUv3
T -	K1OD	K8MP	kSfH	anLQ	3gUw	0S19	Kpjm	Sin1	5iRU	mcJh	Uqog	IXDU	4sOH
U -	Deaz	T1X6	M1wG	cvn4	QTYb	ShxD	Jx3b	UgRQ	LeX5	PkAU	YNyF	Z1vx	4yfx
V -	Lqvv	gav1	KmEd	i9R2	7xCd	ublv	6Okk	GFyT	N8hF	XgdI	uW7u	5CF7	H4Pm
W -	Q830	3Als	KBxV	Cj2I	cAf5	E0HH	u3WG	EOU5	p8Yu	UAF8	dXg6	qeHT	AlFH
X -	dBCR	eZ6B	56yL	e185	TNnO	ANiC	8kSE	0TjK	cRHh	Uf1j	GmQw	KoHK	zT5b
Y -	Hd5H	D6Z5	sICn	zDjy	D2cd	4Eyv	mSsa	Hit2	k3b1	LzF4	19KF	F7hF	NF4K
Z -	iWtb	5PBJ	G2Lm	zJRN	z5S6	45VJ	F0h9	hv06	WEg2	Q92d	MPTT	iNuq	7mhw

A password creation matrix.

To provide a password hint using the previous table, simply choose two random pairs of letters (a total of four letters). For example, let's choose AV and PT. Beginning with the first letter pair, locate the first letter in the left most column and the second letter in the top reference row. Where this column and row intersect on the table is the first half of your password. In our example, AV is Lqqv. Repeating the process using the second pair of letters, PT, you get the second half of our password, kFsH. Combining these two halves our password for encryption/decryption of a message is "LqqvkFsH". However the password hint we send is AVPT.

I use this password table with a small group of friends. We change the table quarterly. This provides reasonably good security for symmetric encryption programs and password hints. The table can also be used to generate passwords for personal use. For example, if I use the password "aBZKJKEh2T8xCkcc" this would be very difficult to guess. Used on a daily basis, it is not all that difficult to remember; no more so than remembering your driver's license number or a credit card number. However, if you were to forget the password, remembering the key word "survival" would allow you to recover the password using the above table and the letter pairs SU RV IV AL.

One additional step you can take to increase the security of your password-hints table is to vary the length of the random sequences in the table itself. In our example, a four-letter password hint always gives an eight-letter password, and an eight-letter hint gives a sixteen-letter password. However, by varying the length of the random sequences in the chart between three and six characters, you can ensure that the length of the password hint does not disclose the password's length.

CODES AND CIPHERS

Codes and ciphers are used to secure our communications, to make them intelligible only to our intended recipients. When we transmit via radio, our message is accessible to anyone capable of monitoring the frequency and mode of communication. (This is also true with methods of communication where we tend not to think of our messages being monitored, such as telephone and e-mail.)

We have discussed computer-based encryption programs, but you may not always have a computer available in order to run these programs. If you are backpacking in the mountains, carrying only a small CW transceiver for communication, a computer-based encryption program will be of no value to you, since you don't have a computer. Still, you might want to send and receive messages that are not intelligible to every person who happens to monitor the transmission. In these cases you can use paper-and-pencil codes and ciphers.

When considering the use of codes and ciphers, we can divide them into two broad categories: (1) those used to enhance communications, and (2) those used to secure communications. Both of these purposes are very valid reasons for using codes and ciphers in our communications.

Examples of codes and ciphers used to enhance communications are Morse code and brevity codes, such as 10-codes or Q-codes. (For example your location is 10-20 in the 10-code and QTH in the Q-code.) These codes do nothing to secure or hide the meaning of a message, but they certainly make communication more effective, especially over difficult communication paths.

Codes and ciphers used to secure communications include the various encryption programs available for use with your computer, one-time pads, or other methods for obscuring the meaning of a message from all but the intended recipient, such as authentication/cipher tables and ops-codes.

There are several very valid reasons for using codes and ciphers for securing communications. It is important to remember that radio transmissions are receivable by anyone with a radio and the ability to tune to the frequency and mode of your signal. While radio operators very often enjoy chatting with strangers and thereby making friends on the air, there are also times when we may want to send personal information to someone without making that information available to everyone monitoring the frequency. For example, you might want to provide your home telephone number to someone by radio, or you might want to provide your exact location to a friend so you can meet, without disclosing that information to others.

Codes and ciphers have a very legitimate place in communications for survival and self-reliance. They are not just for spies, drug dealers, your bookie, or members of organized crime. They are also very important tools for safeguarding the privacy of self-reliant, freedom-seeking individuals. This having been said, it is important to be aware that Big Brother does not like average citizens like us to use codes and ciphers to ensure our personal privacy. The use of codes and ciphers to obscure the meaning of a message might be prohibited when using certain types of communication (such as Amateur Radio). It is important to be aware of the law in these matters, understanding when code and ciphers are permitted and for what purpose they can be used.

One-Time Pads

When discussing codes and ciphers the question will often arise as to whether there are any unbreakable ciphers. The answer is yes! The one-time pad, when properly constructed and properly used, is unbreakable. Unlike other cipher systems that were thought to be unbreakable because of their complexity, the one-time pad can be mathematically proven to be unbreakable!

The key to the security of a one-time pad is a truly random key—and herein lies the problem with one-time pads. It is very difficult for the average person to generate a truly random key of any length. A random string of letters could be generated by a Lotto-type drawing—drawing a ball or tile containing a letter from a constantly mixed bin. Once a letter was drawn and recorded, it would be necessary to immediately return it to the bin so that there was an equal potential that it could be drawn again. This would produce a random string of letters, but it would be extremely slow.

Processes generated by computer programs are not truly random. If one has a copy of the program and knows the key or starting sequence, the program can be used to duplicate the "random" string, thus proving it is not truly random.

Even if it were possible to generate large numbers of truly random keys, you are then faced with the problem of securely distributing those keys to the people who will use them. The keys must be sent by an absolutely secure means. You can't just e-mail them to your friends and expect to have perfectly secure communication thereafter. The moment you send the keys over the Internet they are compromised. Even if you encrypt the keys with some other program (e.g., PGP), the keys are only as secure as the program used to encrypt them. The only really secure way to transmit the keys used in a one-time pad is to pass them in a face-to-face meeting directly to the person with whom you will use those specific keys.

Because of the difficulty of creating truly random keys and securely distributing said keys, the one-time pad is not frequently used among large groups passing large numbers of messages. A one-time pad might be useable by an ambassador at an embassy to pass secure diplomatic messages home, but this same system would be completely unusable by an army on the move sending reports from numerous companies and battalions every day.

Understanding the limitations of the one-time pad, it can be an excellent tool for secure communication for the purposes of survival and self-reliance. We make the assumption that for these purposes, the number of highly secured messages that need to be passed will be few and that those messages will only be passed among a small group of people. This resolves the problem of having to generate and distribute large numbers of keys.

The other problem we discussed was that of generating truly random keys. Slow methods, such as a lottery-type drawing of lettered balls or tiles, the numeric results of a spinning roulette wheel, or the roll of dice are not an overwhelming problem if you are only producing a few one-time pads.

If you have the need to produce a greater number of one-time pads, it is possible to use a pseudo-random string produced by computer. We have already said that a sequence produced by a computer program is not truly random, but there is a concept known as Cryptographically Secure Pseudo-Random Number Generation (CSPRNG). This simply means that, while the process involved is not truly random, it is sufficiently random to provide cryptographic security against all current technology and means of cryptanalysis. An excellent program for CSPRNG of one-time pads can be obtained from the Fourmilab Switzerland Web site at www.fourmilab.ch/onetime/otpjs.html.

A one-time pad is a random (or at least cryptographically secure pseudo-random) string of letters or numbers that is combined with the plaintext of a message in order to produce the ciphertext. There are several ways to combine the random string with the plaintext. One commonly used method is to number the letters of the alphabet in order (i.e., A=1, B=2, C=3 . . . Z=26) and add these numbers to the numbers of the random string of the one-time pad. This is effective but tends to increase the length of the message, because a letter (a single character) is represented by at least a two-digit number, and usually a longer number.

A more effective way of using the one-time pad is to use it in conjunction with a Vigenere table, where a letter from the one-time pad and a letter from the plaintext are associated using the Vigenere table to produce a letter of the cipher text.

When using a one-time pad in conjunction with a Vigenere table, first write out the message you wish to encrypt. Below each letter of the message write the letters of the one-time pad. Then use the Vigenere table to obtain the ciphertext of the message, by recording the letter on the Vigenere table where the plaintext letter and key letter from the one-time pad intersect.

For example, if we wish to encrypt the word "Vigenere" and the key letters from our one-time pad are L E W H Y D I E, we begin by locating the first letter of our plaintext word Vigenere—V—in plaintext alphabet across the top of the Vigenere table, and the first letter of our key from the

Vigenere Table.

PlainText

	A	B	C	D	E	F	G	H	I	J	K	L	M	N	O	P	Q	R	S	T	U	V	W	X	Y	Z
A	A	B	C	D	E	F	G	H	I	J	K	L	M	N	O	P	Q	R	S	T	U	V	W	X	Y	Z
B	B	C	D	E	F	G	H	I	J	K	L	M	N	O	P	Q	R	S	T	U	V	W	X	Y	Z	A
C	C	D	E	F	G	H	I	J	K	L	M	N	O	P	Q	R	S	T	U	V	W	X	Y	Z	A	B
D	D	E	F	G	H	I	J	K	L	M	N	O	P	Q	R	S	T	U	V	W	X	Y	Z	A	B	C
E	E	F	G	H	I	J	K	L	M	N	O	P	Q	R	S	T	U	V	W	X	Y	Z	A	B	C	D
F	F	G	H	I	J	K	L	M	N	O	P	Q	R	S	T	U	V	W	X	Y	Z	A	B	C	D	E
G	G	H	I	J	K	L	M	N	O	P	Q	R	S	T	U	V	W	X	Y	Z	A	B	C	D	E	F
H	H	I	J	K	L	M	N	O	P	Q	R	S	T	U	V	W	X	Y	Z	A	B	C	D	E	F	G
I	I	J	K	L	M	N	O	P	Q	R	S	T	U	V	W	X	Y	Z	A	B	C	D	E	F	G	H
J	J	K	L	M	N	O	P	Q	R	S	T	U	V	W	X	Y	Z	A	B	C	D	E	F	G	H	I
K	K	L	M	N	O	P	Q	R	S	T	U	V	W	X	Y	Z	A	B	C	D	E	F	G	H	I	J
L	L	M	N	O	P	Q	R	S	T	U	V	W	X	Y	Z	A	B	C	D	E	F	G	H	I	J	K
M	M	N	O	P	Q	R	S	T	U	V	W	X	Y	Z	A	B	C	D	E	F	G	H	I	J	K	L
N	N	O	P	Q	R	S	T	U	V	W	X	Y	Z	A	B	C	D	E	F	G	H	I	J	K	L	M
O	O	P	Q	R	S	T	U	V	W	X	Y	Z	A	B	C	D	E	F	G	H	I	J	K	L	M	N
P	P	Q	R	S	T	U	V	W	X	Y	Z	A	B	C	D	E	F	G	H	I	J	K	L	M	N	O
Q	Q	R	S	T	U	V	W	X	Y	Z	A	B	C	D	E	F	G	H	I	J	K	L	M	N	O	P
R	R	S	T	U	V	W	X	Y	Z	A	B	C	D	E	F	G	H	I	J	K	L	M	N	O	P	Q
S	S	T	U	V	W	X	Y	Z	A	B	C	D	E	F	G	H	I	J	K	L	M	N	O	P	Q	R
T	T	U	V	W	X	Y	Z	A	B	C	D	E	F	G	H	I	J	K	L	M	N	O	P	Q	R	S
U	U	V	W	X	Y	Z	A	B	C	D	E	F	G	H	I	J	K	L	M	N	O	P	Q	R	S	T
V	V	W	X	Y	Z	A	B	C	D	E	F	G	H	I	J	K	L	M	N	O	P	Q	R	S	T	U
W	W	X	Y	Z	A	B	C	D	E	F	G	H	I	J	K	L	M	N	O	P	Q	R	S	T	U	V
Y	Y	Z	A	B	C	D	E	F	G	H	I	J	K	L	M	N	O	P	Q	R	S	T	U	V	W	X
Z	Z	A	B	C	D	E	F	G	H	I	J	K	L	M	N	O	P	Q	R	S	T	U	V	W	X	Y

Key

one-time pad in the key alphabet column to the far left of the Vigerene table. V and L intersect on the Vigenere table at the letter G, thus making G the first letter of our cipher text. We then take the next letter of our plaintext (in this case I) and the next letter from our one-time pad (in this case E), giving us the ciphertext letter M. Continuing in this manner until the entire message is encrypted, we find that the word "Vigenere" encrypts to the ciphertext: GMCLLHZI.

To decrypt a message using a one-time pad and Vigenere table, we simply reverse the encryption process. Write the letters of the ciphertext directly below the letter of the one-time pad. Locate the first letter of the one-time pad (in our example, L) in the left-hand key column. Follow this row across until you locate the first letter of the ciphertext (in this case, G). Follow this column to the top of the table and read the plaintext letter from the plaintext row (in this case, V). Simply repeat this procedure for each letter of the ciphertext until you have decrypted the message and can read the plaintext.

Once a page/key of the one-time pad has been used, it must be destroyed immediately. The security of the one-time pad is contained in the security of the key. A one-time pad key must *never* be used a second time. If a one-time pad key is used twice and both messages are intercepted, they can be broken through basic cryptanalysis.

When properly constructed and properly used, the one-time pad provides unbreakable encryption. However, as we have seen there are some difficulties involved in constructing and distributing random keys (the one-time pads themselves). For your most sensitive information, that information that simply must never be compromised, the one-time pad is certainly the best encryption option. However, as we have seen, a less complex (and less secure) method of encryption might serve.

BlackHorse Transposition Cipher

This transposition cipher was used by the U.S. military as late as the 1980s. The cipher can be used to scramble a message of any length (although messages should be kept short for efficient handling).

To use this cipher, a key word is chosen. In this case we will use the keyword "BlackHorse" which is a 10-letter word without any repeating letters. To encrypt a message using this cipher the keyword is written across the top of a page and the letters of the plaintext message are written so that they fall directly below the letters of the keyword.

2	7	1	3	6	5	8	9	10	4
B	L	A	C	K	H	O	R	S	E
M	E	N	A	R	E	O	N	L	Y
T	R	U	L	Y	F	R	E	E	T
O	S	P	E	A	K	T	H	E	I
R	M	I	N	D	I	N	O	P	E
N	D	E	B	A	T	E	W	H	E
N	T	H	E	Y	A	R	E	A	L
S	O	F	R	E	E	T	O	S	P
E	A	K	P	R	I	V	A	T	E
L	Y	W	I	T	H	O	U	T	F
E	A	R	O	F	G	O	V	E	R
N	M	E	N	T	S	U	R	V	E
I	L	L	A	N	C	E	A	N	D
C	E	N	S	O	R	S	H	I	P

"Men are only truly free to speak their mind in open debate when they are also free to speak privately without fear of government surveillance and censorship."

Once the message is written beneath the letters of the keyword, the letters of the keyword itself are numbered in alphabetical order (as shown above).

The letters of the message are then copied from the columns of letters below the keyword, starting with column #1 and ending with column #10. As the letters are copied from the columns they should be written down in three or five letter groups to aid in transmission. So our ciphertext becomes:

NUPIE HFKWR ELNMT ORNNS ELENI CALEN BERPI ONASY TIEEL PEFRE DPEFK ITAEI HGSCR RYADA YERTF TNOER SMDTO AYAML EORTN ERTVO OUESN EHOWE OAUVR AHLEE PHAST TEVNI

If you counted the number of letters in the plaintext message you would see that the complete message consists of 130 letters in the ciphertext. To decrypt this message it is first necessary to write the keyword (in this case BLACKHORSE) across the top of a sheet of paper and number the letters in alphabetical order. Next count the number of letters in the ciphertext (in this case 26 groups of 5 letters for a total of 130 letters). Divide the length of the keyword in letters (10 letters) into the total number of letters in the cipher text (130 letters), giving you the number of letters in each column (in this case 13).

To decrypt this message, it is first necessary to write the keyword (in this case, BLACKHORSE) across the top of a sheet of paper and number the letters in alphabetical order. Next count the number of letters in the ciphertext (in this case, 26 groups of 5 letters for a total of 130 letters). Divide the length of the keyword in letters (10 letters) into the total number of letters in the cipher text (130 letters) yielding the number of letters in each column (in this case, 13).

Now copy the first 13 letters of the ciphertext into a column under the #1 in the keyword (under the letter A in this case). Copy the next 13 letters into column # 2, the next 13 letters into column # 3, and so forth, until all the columns have been filled with the cipher text. Now simply read the message left to right and top to bottom as it has appeared in the block of letters.

While this single transposition provides some degree of security, it is solvable using a digraph attack against a sufficiently long text. To provide a greater degree of security, use a double transposition—follow the above procedure and then scramble the ciphertext using a new keyword. This double transposition would not be easily broken unless a great number of messages (encrypted with the same key pair) were intercepted. The keywords need not be the same length. In fact, it provides greater security if the keywords are of different length.

The weakness of double transposition is that its intended recipient will use the keywords in the wrong order, thus making the message unreadable. However, with reasonable care in encryption and decryption the double-transposition cipher provides moderately good security.

Morse Code – Quick Cipher

The Morse Code – Quick Cipher is a concept developed by the German military in 1918. This cipher, used during World War I, was known as the ADFGVX Cipher. It uses a combination of substitution and transposition to safeguard its plaintext.

Encryption begins with a 6 x 6, containing 36 blocks that are filled with the 26 letters of the alphabet and the numbers 0 to 9, in a random order. Each row and column of the grid is labeled with the letters ADFGVX.

	A	*D*	*F*	*G*	*V*	*X*
A	W	T	3	E	B	2
D	Q	K	9	P	6	V
F	Z	J	G	N	I	Y
G	R	C	L	0	H	5
V	A	M	S	U	X	D
X	4	O	8	1	F	7

To encrypt a message, simply record the letter pairs that correspond to the location of the letters of your message in the grid. Read the first letter of the pair from the left column and the second letter of the pair from the top row. Where this column and row meet is the plaintext letter associated with the cipher-letter pair. For example, the letter pair corresponding to the letter H is GV. Using the above grid, you can encrypt the message Change Freq. To 14070L as GD GV VA FG FF AG XV GA AG DA AD XD XG XA GG XX GG GF.

So far this is nothing more than a simple substitution cipher, substituting letter pairs for single letters and numbers. This offers no security and doubles the length of the message.

Next the cipher pairs are associated with numbers in random order, in this case the numbers 1 – 6 (the number key). A keyword can be used as a mnemonic (as with the BlackHorse cipher) to aid in remembering the order of the number key. In the following example, we have used a number that is 6 digits in length, but a longer or shorter number or keyword could also be used.

4	2	5	3	1	6
G	D	G	V	V	A
F	G	F	F	A	G
X	V	G	A	A	G
D	A	A	D	X	D
X	G	X	A	G	G
X	X	G	G	G	F

The 36 letters of the cipher are written left to right and top to bottom, below the numbers 1 – 6. Finally, three-letter groups are copied from the columns in numerical order. The completed cipher message reads: VAA XGG DGV AGX VFA DAG GFX DXX GFG AXF AGG AGF.

The cipher text is composed of just six letters, which can be easily sent by Morse code. The reason for these particular six letters is that they are dissimilar in Morse code, not likely to be confused during transmission.

Decrypting a message sent using the Morse code – Quick Cipher simply reverses the encryption procedure. Begin by copying the code groups into a grid under the numeric key, filling in each column in numerical order. Then read the letter pairs left to right and top to bottom, decrypting the message using the cipher grid.

The cipher grid is easily constructed by randomly drawing tiles containing the alphabet and numbers 0 through 9 from a container and recording the drawn letter or number onto the grid. The key number can also be chosen for each grid by drawing the numbers 1 through 6 in like manner or by using a keyword and numbering the letters in alphabetical order.

This cipher system is not intended to send long or complex messages. It is, however, useful for sending brief instructions, such as the frequency change in the example above, or perhaps a message such as, "Call me at 202-555-1212."

The advantage to this cipher is that it is easily constructed by anyone. It is specifically intended to secure brief messages sent by Morse code (although it could be used in conjunction with any type of transmission). The cipher provides a fair degree of security to protect such things as telephone numbers, addresses, frequency changes, and like items that you might not wish to send as plaintext by radio. Once a grid and number key have been used, they are simply discarded, with a new grid and number key being used for the following message.

Because you will probably have several of these cipher grids and their number keys prepared, each grid/number pair is given a code word identifying that combination. Thus, the cipher grid and number key used in the previous example might be referred to as "Saturn," while a different grid and number might be "Neptune." In this case, the complete message using the above cipher would be: SATURN, Groups 12, VAA XGG DGV AGX VFA DAG GFX DXX GFG AXF AGG AGF.

An authentication/cipher table consists of the alphabet written in a single column (vertically) in alphabetical order, then each of those letters in the vertical column is followed by the remaining letters of the alphabet (sans that particular letter), repeated in random order in horizontal rows. This provides a table consisting of the alphabet repeated at random 26 times. The scrambled letters of the alphabet are then divided into groups of two to four letters, giving a total of 10 groups (not counting the left-most column). These groups are then numbered zero through nine at the top of the table. The specific numerical division of these groups may differ from one authentication/cipher table to the next. Finally, the alphabet is printed in alphabetical order at the bottom of the authentication/cipher table. This line serves as a user's guide and is not used in authentication.

Authentication/cipher table.

	0	1	2	3	4	5	6	7	8	9
A	HDJY	NCW	GZU	OI	LB	EFQ	RX	TS	KP	UM
B	PNYW	JGM	DEL	XR	SZ	HOQ	CK	FA	UU	TI
C	NXKW	HQT	SUE	DG	UY	ILP	RB	ZO	AJ	FM
D	QALF	NUH	CEY	BO	JW	RPZ	GX	SK	MT	IU
E	RMUD	LSX	OBN	GF	WA	QKZ	PI	CJ	UT	HY
F	BPNG	JYR	UKD	EZ	WX	HQI	CL	SA	MO	TU
G	YUZL	QRC	ONS	TB	KM	FJA	PX	DW	HI	EU
H	WYDG	BMP	CRT	AN	SF	ZUQ	IU	XK	JE	LO
I	QBSO	WDU	JMZ	PX	UR	NEH	YT	AL	CG	FK
J	RKZY	SLB	OUI	AF	DT	CEU	HG	XM	QN	WP
K	TUNB	EWQ	OML	ZJ	AR	XUF	DG	CY	PH	SI
L	NUGR	YAU	EKZ	PH	TC	DMF	IB	QS	WJ	XO
M	JZBO	RFH	SQK	YL	WT	CXP	NU	ID	UG	EA
N	LQFZ	UMR	WOU	IP	GK	SYC	EX	HJ	AB	DT
O	ZGPM	AUQ	UDC	JS	NE	IKT	WL	BF	YX	HR
P	HSGU	DOI	FQR	CX	AL	UTE	NY	ZJ	BW	KM
Q	KUUX	CJI	LZA	NF	YH	PEW	BO	SR	DT	MG
R	NPXS	GLA	UWU	CT	OE	QJF	ZK	HY	DB	IM
S	QGKN	TAI	OEM	DC	LR	UYF	XU	WH	ZP	BJ
T	SFUO	HIM	JXQ	RA	EB	KLP	DC	ZW	UG	NY
U	GDXZ	LTA	BHF	OJ	UQ	WPM	NE	CY	SR	IK
V	QFCB	RKL	ESI	GJ	XD	HYT	PZ	AN	OU	WM
W	HPGO	AJT	LQC	IR	FD	KXM	UE	ZS	UB	NY
X	UUGO	ZER	TKD	JS	HM	BQL	IF	NY	CW	AP
Y	KBOE	ZFH	NQT	IU	LA	SRG	WM	JC	UP	XD
Z	OBUM	KRY	UPL	JC	QF	ESG	AX	IN	TH	WD
^Set Letters										
A	BCDE	FGH	IJK	LM	NO	PQR	ST	UV	WX	YZ

Once you have produced an authentication/cipher table, you can begin to use it to authenticate your communications and validate the authenticity of others with whom you are in communication. Authentication is designed to protect a communications system against the acceptance of fraudulent transmissions. Authentication is done as a challenge and reply.

To prepare an authentication challenge, choose two letters at random. In this example we will choose the letters W and X. Your first letter is always chosen from the left-most column of letters in the authentication table (the set letters). Your second letter is chosen at random from that row of letters.

So . . . having chosen the letters W and X, we issue an authentication challenge as: "Authenticate Whiskey – X-ray." The person receiving the authentication challenge looks at his copy of the authentication/cipher table and follows row W across to the letter X. Directly below the letter X in row W is the letter Q. This becomes the reply to the authentication challenge. Thus, the proper reply to the challenge "Authenticate Whiskey – X-ray" is "I authenticate Quebec."

Having properly responded to the authentication challenge, the replying station will now reverse the challenge (choosing two new random letters) and challenge the other station. Once both stations have properly authenticated, it may be assumed that they are who they claim to be and that the messages being transmitted are valid.

The authentication/cipher table can also be used to self-authenticate messages that are sent in the blind (as a broadcast) or in non-real-time (such as to a PBBS). In this case, you will follow the same procedure of choosing two random letters and identifying the letter in the table directly below the second letter, but if you are not in direct (two-way communication) with the station to receive your message, you simply add the authentication as a three-letter group. For example, if you chose the letters P and V as your initial random pair, you self-authenticate a message by saying, "Authentication is Papa-Victor-X-ray."

Authentication is a useful process to confirm the identity of stations/persons with whom you exchange messages as well as to confirm the validity of messages exchanged. The authentication/cipher table can also (as the name implies) provide a simple cipher to encrypt numbers.

At the top of the authentication/cipher table are the numbers zero through nine, centered above groups of random letters in the table. If you identify which line of the table is being referred to, you can transmit letters representing the numbers at the top of the table. To identify a specific line in the table for encryption, simply choose two random letters (as you did in authentication). You will set the proper line by transmitting those letters to the receiving station, saying, for example: "I Set Echo-November." The receiving station looks at his copy of the authentication/cipher table and now knows that the cipher text will come from line D (Delta).

To send the cipher text after setting the line, you say, "I Send" and then read the cipher letters needed for your message. In this example, working from line D, if we want to send the number 4, we can encrypt it using either the letter J or W since these two letters are in the group below the number 4 in line D. If we encrypt the number 2 we can use the letters C or E or Y, since all of these letters are in the group below the number 2 on line D.

This cipher function can be used to protect numeric elements of messages. It provides a short-term advantage when it is impossible or impractical to secure information to any greater degree. This type of cipher can be used to protect the when, where, and how many in communications that might otherwise be transmitted in an unencrypted format.

For example if you wanted to tell a station to change frequencies to 14067 at 1600 hours, you could send the following message: "Change frequency to I Set: E-N. I Send: L J Q G S at V X A L." Someone hearing this transmission would know that the receiving station would be changing frequencies but would not know what the new frequency was or when the change would occur.

Brevity Lists

The authentication/cipher table is an efficient method of encrypting short groups of numbers; however, used alone it only protects numeric elements of the message. But the authentication/cipher table can also be used with an established list of words and phrases to send more complex messages. This list of words and phrases is numbered and is referred to as a "brevity list." The brevity list can be a simple list of half a dozen instructions, or it can be a more extensive dictionary-like list of words allowing much more detailed and complex messages.

For example, if your brevity list contains an instruction "859 = Change Frequency To #" you can send our above frequency change instruction as "I Set: E-N. I Send: M R I L J Q G S at V X A L."

It is important to remember that a brevity list alone does *not* provide security. The brevity list is a standardized list in alphabetical/numerical order, and once established for a specific group will not change too often. However, when the numbers—indicating words and phrases of the brevity list—are encrypted using the authentication/cipher table (which is changed at least daily), there is a good degree of security applied to your messages.

Appendix I contains a long alphabetical list of numbered words as an established brevity code. This is just the alphabetical listing of a military ops-code with the military-specific terms (e.g., "request air strike at ___") removed from the list and a few new words added. This list, when combined with your own authentication/cipher tables, will allow you to secure your communications from those who might be listening. You will also want to adjust the brevity list to meet your own communications needs. If there are words and phrases that you frequently use in your messages, be sure they are contained in the brevity list. If the list contains words or phrases that you never use, simply remove them from the list to avoid cluttering up your brevity list with nonessential text and to allow for the addition of words and phrases important to your communications.

One limitation of using the numeric brevity list in conjunction with the authentication/cipher table is that you are limited to 1,000 words and phrases (000 – 999) on the brevity list if you maintain the standard three-number groupings.

It is important to keep your messages short when using the authentication/cipher table and brevity list for security. If your encrypted message is longer than 25 or 30 letters, you should choose a new pair of "Set Letters" after each 25 letters. This will add to the security of your encryption.

The authentication/cipher table used for numeric encryption only or in conjunction with a brevity list for more complex message encryption is *not* intended to provide an extremely high level of security. It does provide fair tactical/operational security and is certainly better than sending unencrypted/plain text. The other advantage of using the brevity list is that it (as one would expect with a brevity list) shortens messages, thus making them more easily transmitted.

Matrix Codes and Code Charts

Small codes can be conveniently printed in the form of a matrix or code chart. These matrices or code charts can contain letters, numbers, syllables, and a small number of words. They are very easy to use and provide greater security than simple ciphers.

Once the code chart has been made, it is easily adjusted from one period to the next by changing the coordinates while retaining the contents (i.e., letters, numbers, words) in their same position within the matrix itself.

A very effective way of using a code chart is to have two values in each cell of the matrix. The upper value is a number, letter, syllable, or symbol. The bottom value of each cell is an actual word. One cell in the matrix is used to instruct the user to shift up (to numbers, letters, etc.), and another cell is used to instruct the user to shift down (to words).

A matrix consisting of the complete alphabet along each axis would provide 26 x 26 squares in the matrix, or 676 words. Using two values for each block in the matrix allows you to encrypt up to 1,352 different words, numbers, symbols, syllables, etc. Generally, however, matrix codes and code charts are used to secure a small number of specific messages and instructions. Larger, more complex codes are best designed in a different format.

Ops-Code

The ops-code is a two-part code where three-letter code groups are substituted for words and phrases, similar to the numbers of the brevity list. The brevity list contains a list of words in alphabetical order with the corresponding number groups in numerical order. This is a cryptological weakness and may lead to discovery of certain words through analysis of repetition and patterns of the encoded text.

The ops-code counters this weakness by listing the numbers in random order, or more frequently using random three-letter groups to expand the length of the code. We have seen that using three-digit numbers (000 – 999) limits us to 1,000 possible words and phrases. However, using three-letter groups allows more than 17,000 possible words and phrases (26 x 26 x 26 = 17,576).

An example of the first 10 words of an ops-code's encode and decode sections follows:

<u>ENCODE</u>	<u>DECODE</u>
PEC a	ABE jeep
BNJ abdomen	ABF friend-ly
FHW able/ability	ABL country
TRT abort-s-ed-ing	ACD orange
MMA about	ACG returnee-s
QAF above	ACJ who
TUL abrasion	AED march
WFP accident-al-s	AFH observe-d-r
NVX accomplish-ed	AFR life
CGO accomplishing	AFT radiation

To encode a message using the ops-code, simply look up the word you want to encode in the alphabetical listing of words in the encode section of the ops-code (much like using a dictionary) and record the three-letter code for that word or phrase. To decode a message using the ops-code, just look up the code group in the alphabetical listing of code groups in the decode section of the ops-code and record the word or phrase listed for that code group. A complete ops-code is included in Appendix II. The ops-code is longer than the brevity list. To create the ops-code, we began with the brevity list and continued to add words and phrases. The ops-code can in fact be a short dictionary and should contain any words or phrases likely to be used in your own communications.

Brevity lists and ops-codes take time to create. The brevity list is easiest to create using a word processor with its sort and formatting functions. Simply create a list of words and phrases to be included in your brevity list and use the word processor's sort function to maintain these words and phrases in alphabetical order. Then use the formatting function to number each word or phrase and you have a working brevity list that can easily be maintained and modified to meet your specific or changing needs.

The ops-code uses the same initial list of words and phrases but is more difficult to create. You must now associate each of these words and phrases with a random three-letter code group that does not repeat itself (thus having one code group associated with two different words and phrases).

The main difficulty in creating an ops-code is generating a series of random three-letter groups. This is not an insurmountable problem, however, as there are several random number/letter generators available on the Internet. Additionally, if you have a basic knowledge of programming and scripting, it is simple enough to create programs that will create an ops-code.

It is simple to use the brevity list in conjunction with the authentication/cipher table and the ops-code once they are constructed. They both provide fairly good security of your messages when properly used. There is, however, a weakness in this type of code. That weakness is in using the code to spell out words not found in the code itself. In both the brevity list and the ops-code, there is a number/letter group that corresponds to each individual letter of the alphabet. This allows for spelling, but in doing so, you are in effect using no more than a simple substitution cipher for that word or phrase of your message. Because of this weakness in the code, to the extent that it's practical, you should avoid using the brevity list or ops-code to spell.

Cipher Masking (Steganography)

While sending encrypted messages might make it difficult or impossible for an unintended recipient to read the plaintext of the message, the fact that you are using codes and ciphers might in and of itself draw attention to your communications.

If a message appears to be innocuous, it will likely be ignored, where an obviously encrypted message might draw unwanted attention. To make a message appear innocuous, it is necessary to disguise the fact that it contains encryption. In one of the previous examples of encryption, we ended up with a message containing several code groups. We can, however, place these code groups into a message in such a way as to make them appear innocuous. For example:

Hi Max,

Not much new going on around here. The weather has been kind of rainy, so I have been staying inside and playing with the radios. I have been studying Morse Code, trying to get my speed up.

Here is what I got from the random groups part of the practice exercise:
NUPIE HFKWR ELNMT ORNNS
ELENI CALEN BERPI ONASY
TIEEL PEFRE DPEFK ITAEI
HGSCR RYADA YERTF TNOER
SMDTO AYAML EORTN ERTVO
OUESN EHOWE OAUVR AHLEE
PHAST TEVNI

Could you please check my copy and see if it's accurate.

Well . . . that's about all the news for today. I'll look for you on the air sometime soon.

73s,

Tommi

In this case, the code groups of our message

are all contained in the text, but the remainder of the message serves as a mask to conceal the fact that the contains encrypted text. This type of masking helps slip messages past censors. This example is not particularly complex, but it does serve to illustrate the principle of masking.

Other possibilities for cipher masking include using certain words and phrases in an otherwise innocuous text to relay information. For example, using the names "Bob" and "Alice" in the same sentence might mean make contact on frequency X at midnight. Using the words "looks like rain" anywhere in the message could mean that you believe your communications are being monitored. Innocuous words in a message can have any meaning you assign to them. It's simply a matter of establishing these "code phrases" with those with whom you wish to communicate covertly.

Obscuring Your Communications

When communicating with someone on a regular basis, it might be advisable to take steps to obscure your communications. Remember, all communication on radio can be monitored by anyone capable of receiving the frequency and understanding the mode of communication being used.

If you want radio operators to be able to generally find you on the air, you operate at a regularly reoccurring time on the same frequency. If your conversation is interesting enough, people will sooner or later begin to listen to it. On the other hand, if you don't want your radio conversations regularly monitored, then you will need to take steps to obscure your communications.

The first and perhaps most important step in obscuring your communications is to alter the times and frequencies on which you communicate. If you are communicating on the 10-meter band today, tomorrow you should be on the 17-meter band and the day after that on the 12-meter band. Of course, you will need to ensure that propagation is good enough on all chosen bands/frequencies to allow efficient communication, but this is simply a matter of communications planning. Likewise, don't communicate at the same time every day. If today's contact was made at 1800 hours, then tomorrow make your contact at 1600 hours or 2100 hours. The whole idea here is to avoid setting a pattern that someone can follow to monitor your communications.

The next step in obscuring your communications is to use an uncommon mode of communication. The most common mode of communication is voice. Anyone who monitors your frequency can understand your communication (assuming he speaks your language). The next most common mode of communication is telegraphy (Morse code). If you don't know Morse code, a message sent that way would be obscure to you but clearly understandable by many, many others. However, we can choose less common modes of communication to help obscure our conversations from casual listeners. For example, the sound card/digital modes (such as PSK31) are unintelligible without the software to decode these signals. PSK31 is a very popular mode, but PSK31 activity tends to be concentrated around certain frequencies on each band. A PSK31 signal sent on a frequency not near the common PSK31 frequencies might go unnoticed. However, PSK31 has a unique sound and someone who uses it regularly could accidentally find your signal and tune it in. Choosing a less common sound card/digital mode (such as MT63) and staying clear of the PSK31 frequencies, around which you can often find other digital signals, will certainly help hide your communication from casual listeners. Someone hearing a MT63 signal could simply mistake it for static or background noise and ignore it entirely. Along the same line of changing times and frequencies, you should consider changing communication modes on a regular basis. If you have multiple communications modes (and since most of the sound card/digital modes software is free, you should), try changing modes on a regular basis. Not only does this give you experience using these different modes, but it keeps anyone from finding your communications by scanning for one unique mode of communication.

Blind Transmission Broadcast

Blind Transmission Broadcast (BTB) is a method of communication where messages are transmitted but the receiving stations do not reply to the message at the time of its transmission. This is a common method of communication among intelligence agents and special operations teams. At a specific time, the transmitting station comes on the air sends a message and then goes off the air. Agents in the field know at what time and on what frequencies to expect these broadcasts. They simply tune in to these signals and copy the message. Because there is no two-way communication and the agents in the field are not transmitting themselves, they have very little risk of detection.

Of course, the transmitting station will want to know that the agents in the field have in fact received and understood the messages sent to them. BTB messages should contain an acknowledgment key. This is simply a word or code group to be transmitted by the recipient of the message to indicate that he has received and understood the message. For example the BTB message might end with the instruction "Acknowledge with: Georgia Pine." At a selected time and on a selected frequency, the agent receiving the message sends the acknowledgment if he received and understood the BTB message. For security purposes, the time and frequency for acknowledgment is always arranged in advance. It is NEVER contained in the BTB message itself. A complete acknowledgment might look like this: VVV VVV DE 007 RRR GEORGIA PINE SK. The VVV is sent to alert the receiving station to an incoming message. DE is radio shorthand for "This is," and 007 identifies the sender. RRR indicates the proword "Roger," and, of course, GEORGIA PINE is the acknowledgment key for the message in question. SK indicates the end of the message.

The station transmitting the actual BTB message could be on the air for quite some time, depending on the length and number of messages to be sent, but this station is almost always in a secure location. If agents in the field are sending BTB messages, they do so from a mobile or portable station, transmitting their message and immediately moving away from the transmission site in case their signal was intercepted and radio direction-finding was used to locate the transmitter. The acknowledgment signal is very brief, thus making interception and radio direction-finding very difficult.

RadioGrams—Abbreviated Message Format

Groups handling message traffic, such as the National Traffic System (NTS) or Military Affiliate Radio System (MARS), often find that they are repeatedly sending messages that have the same general content. To aid in the rapid and accurate transmission of these messages, they are standardized into a numbered radiogram list. For example, a common message might be, "Arrived Safely." Using the ARRL numbered radiograms, this message would be ARL64. The message "Medical Emergency Situation Exists Here" is ARL13, and the message "Wish We Could Be Together" is ARL57. Numbered radiograms tend to be broken into two major categories: emergency/relief messages and routine/general greetings messages.

Using this same basic concept, we can establish a standard abbreviated message list for use in survival and self-reliance communications. In setting up this system, we work from the assumption that communication is established at the most basic level—Morse Code. Based on this assumption, we establish a matrix in the same format as the Morse Code Quick Cipher, except in place of letters and numbers to be encrypted, we have 36 standard messages that can be sent with simple trinomes (three-letter groups).

RadioGrams—abbreviated message list.

#	AAA	DDD	FFF	GGG	UUU	XXX
AAA	MSG1	MSG2	MSG3	MSG4	MSG5	MSG6
DDD	MSG7	MSG8	MSG9	MSG10	MSG11	MSG12
FFF	MSG13	MSG14	MSG15	MSG16	MSG17	MSG18
GGG	MSG19	MSG20	MSG21	MSG22	MSG23	MSG24
UUU	MSG25	MSG26	MSG27	MSG28	MSG29	MSG30
XXX	MSG31	MSG32	MSG33	MSG34	MSG35	MSG36

Message 1 – Everything is OK here. There is no need for concern.
Message 2 – Only minor property damage here. Don't be concerned about disaster reports.
Message 3 – Please advise your current condition and if assistance is needed.
Message 4 – Return home (to base) as soon as possible.
Message 5 – I will contact you as soon as possible.
Message 6 – I am moving to new location. Send no further communication.
Message 7 – Advise _____ to standby to provide emergency assistance, instructions, or information.
Message 8 – Establish radio communications net on _____ kHz.
Message 9 – Please contact me as soon as possible.
Message 10 – Medical emergency situation exists here.
Message 11 – Situation here critical.
Message 12 – Property damage in this area severe.
Message 13 – Hazardous conditions here. Do NOT approach this area.
Message 14 – Advise health and welfare status of _____.
Message 15 – Temporarily stranded. Will advise when able to proceed.
Message 16 – Report the extent and type of conditions now existing at your location.
Message 17 – Report at once accessibility and best way to reach your location.
Message 18 – Report at once weather conditions at your location.
Message 19 – Arrived safely at _____.
Message 20 – Arriving at _____. Please arrange to meet me there on (DATE/TIME).
Message 21 – I am in hospital at _____.
Message 22 – Require immediate supply of _____.
Message 23 – Unable to proceed. I require assistance.
Message 24 – Evacuation of this area is imminent. May depart without further notice.
Message 25 – Emergency communications network established. All stations please report current status.
Message 26 – Establishing camp at _____.
Message 27 – Require transportation for _____.
Message 28 – This station is operating on emergency (backup) power.
Message 29 – Severe weather conditions here.
Message 30 – Request you make contact with me

on primary frequency, per SOP.
Message 31 – I have postal mail for you. Please provide your postal address.
Message 32 – Please provide a current e-mail address for _____.
Message 33 – Monitor broadcast on _____.
Message 34 – Require _____ support at my location.
Message 35 – Please provide telephone number where I may call you.
Message 36 – Reserved (designated as needed).

Set up as described, this message system allows survival and self-reliance information to be transmitted in a brief, easily understood format. Each message consists of two trinomes. These messages are often referred to as three-by-three (3X3) messages. To help distinguish these messages from other traffic, we also send the number 33 in conjunction with the two trinomes. For example, to send message #1 ("Everything is OK here. There is no need for concern"), we send 33 AAA AAA. To send message # 26 ("Establishing camp at _____."), we simply read the first trinome from the left column and the second trinome from the top row and send the message 33 VVV DDD.

The primary advantage of the 3X3 messages system is that it can be sent by Morse code under adverse conditions, even when the Morse code operator has only the most rudimentary skill. Because each trinome consist of a single letter repeated three times, and the letters chosen for this system are easily identified in Morse code, there is very little chance of confusion in receipt of these 3X3 messages.

The above message list was created for the purpose of this book but in testing seems to work extremely well. The 3X3 message system can contain any set of standardized messages. As long as everyone participating in the communications network has the same list, you have a highly effective communications system.

CHAPTER 10

Training and Study Options

When it comes to training and study options for radio and personal communications systems, the field is somewhat sparse in most schools and colleges. However, there are a handful of organizations providing top-quality courses and training material for those interested in gaining an in-depth understanding of communications electronics.

Just some of the many study guides available to help you earn your Amateur Radio license.

CIE

Cleveland Institute of Electronics (CIE) is the leading school providing nonresident training in electronics in the United States today. Its programs are detailed, in-depth instruction in a wide range of electronics and computer-related areas. CIE programs also tend to be fairly expensive. The programs are certainly worth the money, but not everyone has $1,000 or more to take these courses.

Recognizing this, CIE has created a number of mini-courses in the $50 to $500 range. These courses include "AC/DC Basic Electronic with Lab," "Soldering Micro Course with Lab," "FCC License Course" (for the GROL exam), and many others.

If you are interested in taking any of these mini-courses, you can request a copy of the CIE Bookstore catalog, where these courses and other references are available for sale.

CIE Bookstore
4781 E. 355th Street
Willoughby, OH 44094
www.ciebookstore.com

COMMAND PRODUCTIONS

Command Productions offers a cassette tape course designed to assist you in preparing for the FCC General Radio Operator's License (GROL) exam. This course consists of six cassette tapes, two large electronics textbooks, workbooks, and instruction on solving the math problems on the FCC GROL exam.

Command Productions
P.O. Box 2824
San Francisco, CA 94126

ARRL

The American Radio Relay League (ARRL) is the leading Amateur Radio organization in the United States and perhaps in the world. ARRL membership is open to any licensed Amateur Radio operator. ARRL has developed a series of online courses dealing specifically with Amateur Radio. They began their online training programs with "Amateur Radio Emergency Communications." The ARRL now offers introductory, intermediate, and advanced "Amateur Radio Emergency Communications;" "Antenna Modeling," "HF Digital Communications," and "Satellite Communications" courses. Each of these courses (and others being developed) is available to ARRL members and nonmembers.

American Radio Relay League
225 Main Street
Newington, CT 06111
www.arrl.org

W5YI

W5YI Group defines its mission as "To assist applicants in obtaining various FCC-issued amateur and commercial radio operator licenses." W5YI has a large amount of training material available to assist you in preparing for these FCC exams.

W5YI Group
P.O. Box 565101
Dallas, TX 75365
www.w5yi.org

CHAPTER 11

Radio Retailers and Resources

Amateur Electronic Supply
5710 W. Good Hope Road
Milwaukee, WI 53223

Associated Radio
8012 Cosner
Overland Park, KS 66204

Communication Headquarters, Inc.
3832 Oleander Drive
Wilmington, NC 28403
e-mail chq@chq-inc.com

C. Crane Company
1001 Main Street
Fortuna, CA 95540

GigaParts, Inc.
P.O. Box 11367
Huntsville, AL 35814
e-mail: hamsales@gigaparts.com

HAM Radio Outlet
933 N. Euclid Street
Anaheim, CA 92801

R&L Electronics
1315 Maple Avenue
Hamilton, OH 45011

Universal Radio, Inc.
6830 Americana Parkway
Reynoldsburg, OH 43068

CHAPTER 12

Suggested Reading

Arland, Richard
ARRL's Low Power Communication
ARRL, 1999

Churchhouse, Robert
Codes and Ciphers
Cambridge, 2002

Ford, Steve
Ham Radio Made Easy
ARRL, 1995

Your VHF Companion
ARRL, 1992–1996

Gibilisco, Stan
Amateur Radio Encyclopedia
TAB Books, 1994

Hogerty, Tom (Editor)
ARRL Repeater Directory 2000/2001
ARRL, 2000

Hudson, Jack, and Jerry Luecke
Basic Communications Electronics
Master Publishing, 1999

Hutchinson, Chuck (Editor)
ARRL Operating Manual, 7th Edition
ARRL 2000

Ingram, Dave
Emergency Survival Communications
Universal Electronics, 1997

Kleinman, Joel, and Zack Lau
QRP Power
ARRL, 1996–1999

Kleinschmidt, Kirk A.
Stealth Amateur Radio
ARRL, 1999

Laster, Clay
Beginner's Handbook of Amateur Radio
McGraw-Hill, 2001

Long, Mark, et al.
The World of CB Radio
Book Publishing, 1987

Straw, R. Dean (Editor)
ARRL Antenna Book, 19th Edition
ARRL, 2000

ARRL Handbook 2000
ARRL, 1999

West, Gordon
Gordon West's General Class
Master Publishing, 1998

Wolfgang, Larry D. (Editor)
ARRL Extra Class License Manual
ARRL 2000–2001

ARRL General Class License Manual
ARRL, 2000

Now You're Talking
ARRL, 2000

Tune in the World with Ham Radio
ARRL, 1989

Wrixon, Fred B.
Codes, Ciphers & Other Cryptic and Clandestine Communication
Black Dog, 1998

APPENDIX I

Brevity List

000 zero
001 one
002 two
003 three
004 four
005 five
006 six
007 seven
008 eight
009 nine
010 a
011 abdomen
012 able/ability
013 abort-s-ed-ing
014 above
015 accident-al-s
016 accomplish-ed
017 accomplishing
018 acknowledge-d-ment
019 act-ion-ing-s
020 act-ive-ivity-s
021 add-ed-itional
022 add-ing-s
023 adjacent
024 administration
025 administrative
055 area-s
056 arm-s-ed
057 army-ies
058 around
059 arrange-d-ing
060 arrange-ment
061 arrive-al-d-ing
062 asap
063 ask
064 assault
065 assemble-d-ing
066 assist-ance
067 assist-ed-ing
068 at (once)
069 attach-ed-ing
070 attach-ment
071 attack-ed-ng-s
072 August
073 authorization
074 authorize-d-ng
075 automatic
076 automobile
077 available
078 aviator/pilot
079 axis (of)
080 azimuth (of)
110 booby trap
111 book
112 boundary-ies
113 box
114 boy
115 break/broken
116 breakfast
117 breast
118 bridge-d-ing-s
119 bug(s)
120 building-s
121 bulldozer
122 bunker-s
123 burst-ing-s
124 by
125 bypass-ed-ing-s
126 c
127 cable-s
128 cache'
129 caliber
130 call
131 camouflage-d
132 camouflage-ing
133 camp
134 canal-s
135 cancel-led-s

026 admit
027 advance-d-ing
028 age
029 agent-s
030 agree
031 air (fld/port)
032 aircraft
033 ALE
034 alert-ed-ing
035 alive
036 all
037 alley
038 alternate
039 altitude
040 am in pos (to)
041 ambulance
042 ammunition
043 amplitude modulation
044 amputate
045 answer-ed-ing-s
046 antenna
047 anti-
048 apply
049 approach-ed-s
050 approach-ing
051 approve-al-d-ing
052 approximate-ly
053 April
054 are/is
081 b
082 back-ed-ing
083 bad
084 band
085 bank
086 barbed wire
087 barometer
088 barrage-s
089 barrier
090 base-d-s-ing
091 battery-ies
092 battle-s
093 baud
094 be-en-ing
095 beach-ed-es-ing
096 before
097 begin
098 below
099 between
100 beyond
101 bike
102 black
103 bleed
104 block-ed-ing
105 blow-ew-own
106 blow-ing (up)
107 blue
108 boat-s
109 bomb-ed-r-ing-s
136 capable-ity
137 capacity-ies
138 capsule
139 capture-d-ing
140 car
141 carry-ied-iers
142 casualty-ies
143 cave
144 CB
145 cease-s-d-ing
146 challenge-d-ing
147 change to/from ciphe
148 channel
149 charge-d-s
150 check-ed-ing-s
151 chemical-s
152 child/children
153 church
154 civilian-s
155 clarification(of)
156 class-ify
157 classification
158 clean-ed-ing
159 clear-ed-ance
160 clearing-s
161 close-d-ing
162 cloth-es-ing
163 coast-s-al
164 collapse
165 collect-ed-ion
166 collect-ing-s
167 column-s
168 combat
169 come-ing
170 comma
171 command-er-ed-ing-s
172 commit-ment
173 commit-ted-ing
174 communicate-d-ions-s
175 company
176 complete-d-ing
177 complete-ion
178 compound
179 compromise-d-ing
180 computer
181 concentrate-d-ing
182 concentrate-s-ion
183 condition-ed-ing
220 CW
221 d
222 damage-d-ing
223 danger
224 darkness
225 day-s/date-s
226 deadline-d-ing
227 death
228 December
229 deception
230 decontaminate-d-ion
231 decrease-d-ing
232 deer
233 defend-ed-ing-s
234 defense-ive
235 degree-s
236 delay-ed-ing-s
237 delete-d-ion
238 deliver-ed-ing
275 drop-ping-ped
276 DTMF
277 due
278 dump-ed-ing
279 during
280 e
281 e-mail
282 east-ern (of)
283 ed
284 effect-ed-ing
285 effect-ive-s
286 electric-al-ity
287 element-s
288 elevate-d-ion
289 eliminate-ed-ing-s
290 elk
291 embank-ment
292 emergency
293 embrace-d-ing

184 conduct-ed-ing
185 confidential
186 connect-ed-ing
187 connect-ion
188 conserve-d
189 conservation
190 consolidate-d
191 consolidate-ing
192 consolidation
193 contact-ed-s
194 contact-ing
195 contaminate-d
196 contaminate-ion-ing
197 continue-ation
198 continue-d-ing
199 control-ed-ing-s
200 conversation
201 convoy-s
202 cook
203 coord-s-d-ing
204 corps
205 correct (me)
206 correct-ed-ing
207 counterattack-ed-s
208 counterattack-ing
209 counterpart
210 cover-ed-ing
211 craft
212 crane
213 critical-ly
214 cross-ed-ing
215 crossroad-s
216 crushed
217 CTCSS
218 culvert
219 cut-off
239 demolition
240 demonstrate-d-ing
241 demonstrate-d-s-ion
242 deny-ied-ies
243 depart-ed-ing-ure
244 deploy-ed-ing
245 deploy-ment-s
246 depot
247 deputy-ies
248 designate-d
249 designate-ion
250 desk
251 destination
252 destroy-ed-ing-s
253 detach-ed-es
254 detach-ment-s
255 determine
256 diesel
257 dig/dug (in)
258 dinner
259 direct-ed-ion-ing
260 disable-d-ing
261 disapprove-d-al
262 disperse-d-ing-al
263 displace-d-ing
264 displace-ment
265 dispose-d-al
266 disregard-ed-ing
267 distance-s
268 district-s
269 ditch
270 divert-ed-ion
271 divide
272 division-s
273 dog
274 down-ed
294 emplace-ment
295 employ-ed-ing
296 en
297 encounter-ed-ing
298 encrypt-ed-tion
299 end-ed-ing
300 enemy
301 engineer-ed-ing-s
302 enlist-ed-ing
303 entrench-ed
304 envelop-ed-ing
305 envelope
306 equip-ing-s
307 er
308 error
309 es
310 escape-d-ing
311 establish-ed-ing
312 estab-ed contact at(
313 estimate-d-ing
314 estimate-ion
315 et
316 eta your loc
317 evac-ed-ing-ion
318 evasion
319 execute-d-ing
320 execute-ion-s
321 exercise
322 exhaust-ed-ing
323 expect-ed-ing
324 expect-ation-s
325 expedite-d-ing
326 explode-d-s
327 exploit-ed-ing
328 explosive-s
329 expose-d-ing
330 extend-ed-ing-ive
331 extend-sion
332 extend-extent
333 extra
334 extract-s-ed-ing
335 f
336 facility
337 fade-ed-ing
338 fail-ed-ing
339 fallout
340 family
341 far
385 future
386 g
387 gallon-s
388 gas-ing-ed
389 gasoline
390 gate
391 general-ize
392 generator(set)
393 get-ting
394 girl
395 give-n-ing
396 GMRS
440 ies
441 if
442 ill-ness
443 immediately
444 imminent
445 immobile-ize-d-ing
446 impact
447 impassable
448 impossible
449 improve-d-ing
450 improve-ment
451 in

342 fatal
343 favor-s-ed
344 favor-able-ing
345 February
346 feet/foot
347 FEMA
348 female
349 FeldHell
350 field-s
351 fight-er-ing
352 final
353 fire-d-ing-s
354 fish
355 flame-d-ing-s
356 flank-ed-ing-s
357 flare-d-s
358 fled
359 flesh
360 flexible-ity
361 flight
362 follow-ed-ing
363 food
364 for
365 force-s-d-ing
366 ford-able
367 fog
368 forearm
369 forest-s
370 fortify-ied
371 fortify-ication
372 forward-ed-ing
373 found
374 fracture
375 frag
376 frequency-ies
377 frequency modulation
378 Friday
379 friend-ly
380 from
381 front-age-ally
382 FRS
383 fuel
384 fuse-d
397 go-ing(to)
398 good
399 government
400 GPS
401 grade-d-r-ing
402 green
403 grid(coord)
404 ground-s-ed
405 group
406 guard-ed-ing-s
407 guide-d-s-ing
408 guide-ance
409 gun-s
410 h
411 had
412 halt-ed
413 Ham/ham
414 hand
415 harass-ed-ing
416 harass-ment
417 have/has
418 head-ed-ing
419 headquarters
420 hear-ed
421 heavy
422 helicopter-s
423 here
424 high-er-est
425 highway-s
426 hill/mt/rise-s
427 hit-ting
428 hold-er-ing
429 home
430 hospital-s
431 hostile
432 hour-s-ly
433 how
434 hundred
435 i
436 ice
437 identify-ed
438 identification
439 ied
452 in-accordance-with
453 inadequate
454 inch-es
455 incident-al-s
456 increase-d-g-s
457 indigenous
458 indirect
459 individual
460 infiltrate-d-r-s
461 infiltrate-g-on
462 inflict
463 inform-ed-ation
464 ing
465 initial-ed-ly
466 insert-s-ed-ing
467 install-ed-ation
468 instrumental
469 intact
470 intend-ed-tion
471 intercept-ed-or
472 intercept-ion-s
473 interfere-d-ing
474 interfere-nce
475 internal
476 interrogate-d
477 interrogate-ion
478 intersection
479 interval
480 isolate-d-ing
481 issue-d-ing-s
482 item-s
483 j
484 January
485 jam-med-ing
486 jeep
487 join-ed-ing
488 joint
489 July
490 jump-ed-ing(off)
491 junction-s(road)
492 June
493 k
494 keep-s-ing
495 kilometer-s
496 know-n-ing
497 l
498 labor-ed-ing
499 lacerate
550 max(range)
551 May
552 meal
553 means
554 measure-d
605 none
606 north-ern(of)
607 not
608 not later than
609 nothing

500 land-ed-ing
501 large
502 last
503 launch-ed-er-ing
504 law
505 lay/laid
506 lead-er-ing
507 leaflet-s
508 leave/left
509 left(of)
510 leg
511 letter
512 level
513 liaison
514 life
515 light-er-ing-d-s
516 limit-ed-ing-s
517 line-d-s(of)
518 listen-ed-ing
519 litter-s
520 load-s-ed
521 local-ize-d
522 locate-d-ing-s
523 locate-ion(of)
524 log
525 logistic-al-s
526 look(for)
527 loss-es-lost
528 loudspeaker
529 low-er-est
530 lower side band
531 lunch
532 ly
533 m
534 machinegun-s
535 magnetic
536 main
537 main body
538 maintain-ing-s
539 maintain-ed
540 maintenance
541 make-ing/made
542 male
543 maneuver-ing-s
544 man-ned-ning
545 many
546 map-ped-ping-s
547 March/march
555 measure-ment
556 mechanic-al
557 mechanize-d
558 medivac needed
559 medical
560 medium
561 meet-ing
562 meet (me)(at)
563 message-s
564 messenger
565 meter-s
566 MFSK
567 mile-s
568 mine-d(field)
569 minimum
570 minim-ize-d
571 minor
572 minute-s
573 miss-ed-ing
574 missile-s
575 missing in act
576 mission-s
577 mobile-ity
578 modulation
579 Monday
580 money
581 moon
582 morale
583 morning
584 more/most
585 Morse
586 motor-ized
587 mountain-ous
588 mount-ed-ing
589 move-d-ing
590 move-ment
591 multiple
592 my
593 my location is
594 n
595 narrow
596 near
597 need-ed-ing
598 net-ted-ting
599 neutral-ize
600 neutralize-d-ing
601 new
602 night-s
610 notify-ied-ies
611 November
612 now
613 nuclear
614 numbers
615 o
616 objective-s
617 observe-d-r
618 observe-s-ing
619 observation(post)
620 obstacle-s
621 occupy-ation
622 occupy-ed-ing
623 October
624 odor
625 of
626 offense-ive
627 officer-s
628 oil
629 on
630 only
631 open-ed-ing
632 operate-ion-s
633 operate-s-al
634 opportunity-s
635 or
636 orange
637 order-ed-ing-s
638 ordnance
639 organize-ation
640 organize-d
641 orient-ing-ation
642 origin-al-ate
643 other
644 our
645 out(of)
646 outpost
647 over-age
648 overlay
649 over-run
650 p
651 packet
652 panel-s
653 paratroop-s
654 part-s
655 party-ies
656 pass-es-ed
657 pass-able

548 match-ed-es-ing
549 market
660 past
661 patrol-ed-ing-s
662 penetrate-d-ing
663 penetrate-ion
664 people
665 per
666 percent-age
667 perimeter
668 period
669 permission
670 permit-ed-ing
671 phase
672 photograph-ed-er-ing
673 pick up
674 piece-d-ing-s
675 pipe line
676 pistol
677 place-d-ing
678 plan-s-ed-ing
679 point-s-ed-ing
680 police
681 platoons
682 port(harbor)
683 position-s-ed
684 possible
685 postpone-d-s
686 pospone-ing
687 post-s
688 post office
689 pound-s
690 prepare-d-ing-s
691 preparation
692 prevent
693 priest
694 primary
695 priority
696 prize-s of war
697 probable
698 proceed-ed-ing
699 progress-ed-ing
700 protect-ed-s-ing
701 protect-ed-ion
702 protect-ive
703 provide-d-ing
704 provide-sion
705 province
603 no/non/neg
604 no fire line
715 radiological
716 raid-er-ing-s
717 railhead
718 railroad-s
719 rain
720 range
721 rate
722 ration-s-ed
723 reach-ed-s-ing
724 ready
725 rear
726 receive-r-d-ing
727 recognize-d-ing
728 recon
729 recover-ed-y
730 red
731 reduce-ing-tion
732 reference-d-s
733 reference (to)
734 refuge
735 refugee(s)
736 regiment-ed-s
737 registration
738 regroup-ed-ng
739 reinforce-d-s-ing
740 reinforce-ment-s
741 relay-ed-ing
742 release-d-ing-s
743 relief/relieve
744 remain-ing
745 rendezvous
746 reorganize-d-ation-s
747 repair-ed-ing
748 repeat-ed-ing
749 repel-ed-ing-s
750 replace-d-ing
751 replace-ment
752 report-ed-ing
753 report-s(to)
754 repulse-d-ing
755 request-ed-ing-s
756 require-d-ing
757 require-ment
758 rescue
759 reserve-d-ing-s
760 resist-ed-ance
658 pass-age-ing
659 password
770 rifle-s
771 right(of)
772 riot-s-ed
773 river-stream-s
774 road-s
775 round-s
776 route-s-d-ing
777 RTTY
778 rucksack
779 ruin-s-ed
780 s
781 safe-r-ty
782 Saturday
783 schedule-d-ing-s
784 school
785 scout-ed-ing-s
786 screen-d-ing-s
787 search-ed-ing-es
788 second-s-ary
789 secret
790 section-s
791 sector-s
792 secure-d-ing
793 secure-ity-s
794 shelter
795 sick
796 sieze-d-ing
797 sieze-ure-s
798 send-ing/sent
799 separate-d-ion
800 September
801 sequence
802 service-d-able
803 shell-s-ing-ed
804 shoot-ing/shot
805 short-ed
806 shoulder
807 sight-ed-ing
808 sign-ed-ing
809 signal-ed-ing
810 silent-ly-ce
811 simulate-d-ing
812 site-s
813 sit/sitrep
814 slope-s-ing-d
815 slow-ed-ing

706 PSK
707 pursue-d-ing-t
708 q
709 question-ed
710 r
711 rad(rad/hr)
712 radar
713 radiation
714 radio-s
761 resist-ing
762 rest
763 restore-s-ed
764 restrict-ion
765 restrict-ive
766 resupply-ed-s-ing
767 result-s-ed-ing
768 return-ed-ing
769 ridge-d-s
816 small-er-est
817 smoke-d-ing-s
818 snow
819 soldier
820 soon-er-est
821 sorry
822 sortie-s
823 south-ern(of)
824 speak
825 spearhead
826 sporadic
827 spot-ted
828 squad-s
829 stand-ing(by)
830 start-ed-ing
831 station-ed-s
832 status(report)
833 stay-ed-ing
834 stop-ped-ing
835 storm
836 street
837 strength-en
838 strike/struck
839 strong-er-est
840 storm-ing-(s)
841 submit-ed-ing
842 success-ful-ly
843 Sunday
844 sunrise
845 sunset
846 superior-ity
847 supplement-ed-ary
848 supply-ed-s-ing
849 support-s-ed-ing
850 surface-d-ing
851 surgery
852 surround-ed-ing
853 survey-ed-s
854 survey-lance
855 suspect-ed-ing
856 swallow-ed(s)
857 switch-ed-ing
858 swollen
859 t
860 tactic-s-al
861 take-ing/took
862 talk
863 tank-er-s
880 these
881 this
882 thousand
883 THROB
884 through
885 thunder
886 Thursday
887 tide
888 time-d-ing
889 tion
890 to(be)
891 today
892 tomorrow
893 ton-s-age
894 tone
895 tonight
896 top secret
897 total
898 tow-ed-ing
899 toward
900 town/village-s
901 trace-r
902 trak-ed-ing
903 traffic
904 trail-er-ed-ing-s
905 train-ed-ing-s
906 transmit-ed
907 transmission
908 transport-ed
909 transport-ing-s
910 transportation
911 trapped
912 truck-ed-s-ing
913 Tuesday
914 turn-ed-ing
915 u
916 unable
917 unauthorized
918 under/undercover
935 verification
936 vertical
937 very
938 violate-s-d
939 violet
940 visible-ity
941 voltage
942 vulnerable
943 w
944 wagon
945 wait-ed-ing
946 war
947 warehouse
948 was/were
949 water
950 waypoint
951 we/us
952 weak-en-ness
953 weapon-s
954 weather
955 week-s
956 Wednesday
957 west-ern(of)
958 what
959 wheel-ed
960 when
961 where
962 which
963 white
964 who
965 wife
966 will
967 wind
968 WinLink
969 wire-s-d-ing
970 with
971 withdraw-ing-s
972 wood-s-ed
973 work-s-ed-ing

864 tank destroyer-s
865 target-s
866 task(force)
867 taxi
868 team-s-ed
869 ted
870 teen
871 telephone-s
872 television
873 temple
874 tentative
875 terminate-d-ing
876 terrain
877 territory
878 terrorists
879 there
919 understand
920 understood
921 unidentified
922 unit-ed-es-ing
923 until(further notice
924 unusual
925 unarmed
926 unharmed
927 up
928 upper side band
929 us
930 use-d-ing
931 v
932 valley
933 vehicle-s
934 verify-ied
974 wound-ed-s
975 wreck-ed-s
976 wrist
977 wrong
978 x
979 x-mit
980 x-ray
981 y
982 yard-s
983 yellow
984 yes
985 yesterday
986 you-r
987 z
988 zap
989 zone-s-in

APPENDIX II

Ops-Code

Ops-Code TEXT TO CODE Serial No. 001 Page 1

RSU (SPACE / NULL)
LIH a
LSY abdomen
CEQ able/ability
AFK abort-s-ed-ing
WYY about
WNC above
ONA accident-al-s
AVF accomplish-ed
KYH accomplishing
QAH ache
MAY acknowledge-d-ment
XWF act-ion-ing-s
PVC act-ive-ivity-s
PSA add-ed-itional
DBO add-ing-s
EMI adjacent
CSH administration
CCD administrative
GVA admit
KFE advance-d-ing
XUF after
RUS age
LWF approach-ing
BNJ approve-al-d-ing
PMT approximate-ly
ADY April
XFY are/is
QEP area-s
APK arm-s-ed
PGB army-ies
XVY around
QNT arrange-d-ing
HSN arrange-ment
CIL arrive-al-d-ing
GOU arrow
PQA article
AAS as
JML asap
BPX ask
LID assault
UAR assemble-d-ing
CCI assist-ance
SQG assist-ed-ing
QMD at
HME at (once)
IFW bed
WVE been
MHN beer
JVM before
RPU begin
UCK below
SBO belt
GXU between
YYY beyond
BSV bike
RWI binoculars
XKJ biological
FRS bit
ADD bite
HHC bivouac-ed-ing
YRX black
UMF bleed
CGI blister
ARH block-ed-ing
SIX blow-en
OTL blow-ing (up)
GHD blue
YDL boat-s

HLV agent-s
IEU agree
RBR air (fld/port)
JBI airborne DF platform
BPS aircraft
BFF airdrop-s
NSM airmobile
TRM alarm
WRM ALE
PCQ alert-ed-ing
VJU alive
JWS all
CFQ alley
KGD alt supply pt
ENU alternate
VLU alternate commo
AJQ altitude
XTG am
AJS am in pos (to)
VEQ amateur
BPP ambulance
UVJ ambush-ed-ing
LII ammunition
VVI amplitude modulation
HTP amputate
EOF an
LNJ and
WFT answer-d-ing-s
QHB antenna
BTJ anti-
RGL apply
BKF approach-ed-s

CJA atlantic
CCY attach-ed-ing
IQN attach-ment
QNN attack-ed-ng-s
JKC August
DUM authorization
MAV authorize-d-ng
NUR automatic
YYL automobile
QNS available
DMR aviator/pilot
TBO axe
VXT axis (of)
UKR azimuth (of)
DJE b
GPF back-ed-ing
BNP bad
VTN ball
KGR band
VNS bank
VEA barbed wire
FBN barometer
AME barrage-s
UBL barrier
TUQ base-d-s-ing
BDB battalion-s
RWA battery charger
ATR battery-ies
TON battle-s
NLF baud
VNB be-en-ing
MXI beach-ed-s-ing

IAE bomb-d-r-ng-s
RIL bone
BVF boobytrap
OWI book
EDT boots
AVL border
FDV boundary-ies
THX bow
TLH box
LDH boy
BTB break/broken
PDU breakfast
PTK breast
OPK bridge-d-ing-s
AYA brother
RBT bug(s)
XME building-s
FRE bulldozer
UVC bunker-s
DAW burst-ing-s
WRT but
CDA by
YJM bypass-d-ing-s
AFL c
XBG cable-s
MDA cache'
KTD caliber
KHV call
UCV camera
JNU camouflage-d
XYV camouflage-ing
YBS camp

Ops-Code TEXT TO CODE   Serial No. 001   Page 2

VDW can
MDT canal-s
BUJ cancel-led-s
QYA candle
SAP capable-ity
IPF capacity-ies
JIA capsule
CDR capture-d-ing
CBR car
OFP carry-ied-iers
YOR casualty-ies

QXH concentrate-d-ing
JTA concentrate-s-ion
VGV condition-d-ing
RBP conduct-ed-ing
ASH confidential
ENO connect-ed-ing
CTX connect-ion
FWJ conservation
ACT conserve-d
KOO consolidate-d
WLY consolidate-ng

BOC decontaminate-d-ion
CEV decrease-d-ing
LAO deer
JKT defend-ed-g-s
IJW defense-sive
WVH degree-s
AUQ delay-ed-ing-s
WVK delete-d-ion
GBT deliver-ed-ing
YIC demolition
YMQ demonstrate-d-ing

VPN cave
JKE CB radio
GPV cease-s-d-ing
RWH challenge-d-ing
GDM chance
RQP change
FLB change to/from ciphe
QMS channel
PYE charge-d-s
EIL check-d-ing-s
QTQ checkpoint
JJK chemical-s
NON child/children
FPN church
UNU circle
BLY city
RTP civilian-s
WFC clarification(of)
JLY class-ify
TKW classification
MQE clean-ed-ing
AEX clear-ed-ance
BMA clearing-s
TAT close-d-ing
NLJ cloth-es-ing
LNA coast-s-al
YXG coffee
KEM collapse
AKO collect-ed-ion
TAB collect-ing-s
NKX column-s
EII combat
GWA come-ing
JKO comma
KCU command-er-ed-ing-s
CPC commit-ment
NHT commit-ted-ing
LAG communicate-d-ions-s
CJL company
EDW complete-d-ing
PKN complete-ion
GQK compound
VKC compromise-d-ing
EKY computer
TTM consolidation
PKE contact-ed-s
MEQ contact-ing
OFX contaminate-d
NKD contaminate-ion-ing
NCM continue-ation
ITD continue-d-ing
NHI control-d-ing-s
UFW conversation
DVX convoy-s
KKO cook
OHB coord-s-d-ing
EQG corporation
UGC corps
YGA correct (me)
UWU correct-ed-ing
APN cost
LNN could
CTW counterattack-ed-s
HHX counterattack-ing
YLI counterpart
UHM cover-ed-ing
GHK craft
BFH crane
IXN critical-ly
LWY cross-ed-ing
CDL crossroad-s
XYY crushed
OYY crystal
TOF CTCSS
GQN culvert
CMG cup
IWG cut-off
KPM CW
NRY d
KMB damage-d-ing
VKL danger
BUK dangerous
VDJ darkness
RVX day-s/date-s
CVV deadline-d-ing
PXF death
QTB December
MFL deception
SPR demonstrate-d-s-ion
QKQ deny-ied-ies
PHF depart-d-g-ure
DRR deploy-ed-ing
QTF deploy-ment-s
CQQ depot
WYC deputy-ies
XAD designate-d
CAU designate-ion
RNI desk
KDQ destination
NOH destroy-d-ing-s
LEP detach-ed-es
BPV detach-ment-s
MTN determine
ONT diesel
IFU dig/dug (in)
PJK dinner
BFR direct-d-on-ng
OLV direction finding
QOW disable-d-ing
SBE disapprove-d-al
PQJ disperse-d-g-l
SMN displace-d-ing
WNW displace-ment
CUW dispose-d-al
GGW disregard-d-ing
ISL distance-s
LJK district-s
EDC ditch
LAD divert-ed-ion
QSF divide
SVW division-s
RCL do
NFC dog
RUN down-ed
NCF downed acft
UGJ drill
KIE drip
CNK drive
YVP driveway
WDX drop-ping-ped
FBP drug-s
MDV DTMF

Ops-Code TEXT TO CODE   Serial No. 001   Page 3

XSX due
VHM dump-ed-ing
NYL during
XVE e
LWB e-mail
MYY east-ern (of)
PYN ed
YQC effect-ed-ing
HDF effect-ive-s
CPL eight
LRB electric-al-ity
UUL element-s
KCQ elevate-d-ion
QPW eliminate-ed-ing-s
LFC elk
FHG embank-ment
BDI embrace-d-ing
EVE emergency
UCB emplace-ment
PWE employ-ed-ing
HFM en
QEY en-ed-ing
FEQ encounter-d-ing
PDV encrypt-ed-tion
DBP enemy
IIB engine
LDV engineer-d-ing-s
ARV enlist-ed-ing
VHP entrench-ed
SOF envelop-ed-ing
PRA envelope
NDM equip-ing-s
BWD er
WGE error
QAO es
BPJ escape-d-ing
VSD estab-ed contact at
YED establish-d-ng
UVO estimate-d-ing
AGD estimate-ion
ABO et
TSG eta your loc
WDK etd
YEB evac-d-ing-ion
AGO evasion

PIN expose-d-ing
CUA extend-d-g-ive
AYH extend-extent
OTF extend-sion
UHG extra
NYT extract-s-d-ing
JYE f
OMU facility
AHW fade-ed-ing
DKU fail-ed-ing
SVI fallout
VWE family
AKP far
UPQ fatal
NLN father
RKF favor-able-ing
JBY favor-s-ed
VEV February
KSJ feet/foot
SPG FeldHell
DUJ FEMA
OFT female
PAO field-s
BQL fight-er-ing
BPI final
FNU fire-d-ing-s
UKM fish
GML five
RNT flame-d-ing-s
OYT flank-ed-ing-s
QKT flare-d-s
XVI fled
NQM flesh
RIF flexible-ity
UTF flight
ANG foe
BCY fog
MOS fold-ing
XQQ follow-ed-ing
XIV food
FFI for
MRS force-s-d-ing
WSY ford-able
VTO forearm
ONV forest-s

ERM frequency-ies
KKG Friday
GMS friend-ly
MTV from
SEF front-age-ally
ETL FRS
AAR fruit
MHQ fuel
IGQ fuse-d
MNM future
WXW g
QRC gallon-s
NML gas-ing-ed
ICX gasoline
CND gate
LFJ general-ize
WYV generator(set)
PMV get-ting
PDO girl
XKR give-n-ing
OOT glass
PXW GMRS
MKV go-ing(to)
FAU good
CWC government
KQU GPS
GLQ grade-d-r-ing
AXK great
VJI green
QGU grid(coord)
DBG ground-s-ed
ESY group
PVF guard-ed-ing-s
RTB guide-ance
WOM guide-d-s-ing
VBV gun-s
GLG gunpowder
ACB h
JYP had
WMF halt-ed
XPI Ham/ham
PBO hammer
SOD hand
CBN handbook
BRF handi-talkie

TYV execute-d-ing
YCN execute-ion-s
PKI exercise
RGQ exhaust-ed-ing
CKN expect-ation-s
ILR expect-ed-ing
UVW expedite-d-ing
VXN explode-d-s
CVS exploit-ed-ing
GHY explosive-s

WWU fortify-ication
KCB fortify-ied
GNL forward-ed-ing
RQK found
OSM four
DTJ fracture
LGA frag
YWR freedom
DUA freeze (ing)
OEU frequency modulation

KPV harass-ed-ing
KGH harass-ment
MVY hate
RVE have/has
MGO he
YSV head-ed-ing
DGP headphones
BPG headquarters
HBF hear-ed
NSG heavy

Ops-Code TEXT TO CODE Serial No. 001 Page 4

RFL held
LRI helicopter-s
RXG her
JGX here
JCY hidden
XHM hide
JOE high speed burst
JYM high-er-est
KIG highway-s
GQM hill/mt/rise-s
DYG him
BBF his
UWT hit-ting
CUJ hold-er-ing
EIY hole
ASX home
QXV hook
SDR hospital-s
GAI hostile
AQX hour-s-ly
VVB how
EGB hundred
CJJ i
BMO ice
BRR identification
VSP identify-ed
AQN IDY
BEJ ied
DPX ies
NDR if
WJC ill-ness
FUN immediately
UGQ imminent

UXE ing
UUT initial-ed-ly
BVN insect
HVY insert-s-ed-ing
TBX install-d-ation
TFB instrumental
FWK intact
KSL intend-ed-tion
SNN intercept-d-or
UJJ intercept-ion-s
BFA interfere-d-ing
BDO interfere-nce
FEM internal
AQB interrogate-d
MGS interrogate-ion
ASL intersection
BWC interval
AFB is
FLH isolate-d-ing
FMX issue-d-ing-s
HLL it
BOM item-s
SXC j
RQM jacket
RYE jam-med-ing
ONU January
EOM jeep
SMH job
DQR join-ed-ing
VEY joint
OJQ July
NMO jump-ed-g(off)
KGE junction-s(road)

VRU last
VAQ lateral
UBP launch-d-r-ing
ECB law
KKN lay/laid
HTI lead-er-ing
CHQ leaflet-s
IQB least
DTS leave/left
MUF left(of)
PUV leg
EQD letter
OPC level
BRC liaison
BBC library
SEA life
BIR light-er-ng-d-s
DLE like-s
CIC limit-ed-ing-s
QLP line-d-s(of)
GYU list
MUU listen-ed-ing
EQX litter-s
DRS load-s-ed
EKQ local-ize-d
TMF locate-d-ing-s
GNY locate-ion(of)
DDA lock
NBT log
SVF logistic-al-s
KOU look(for)
YII loss-es-lost
YEL loudspeaker

NJL immobile-ize-d-ing
OGG impact
AXI impassable
PWR impossible
NFY improve-d-ing
LET improve-ment
PNX improvised
NGB in
LLD in-accordance-with
FKG inadequate
MMS inch-es
MIQ incident-al-s
ING increase-d-g-s
LWV indigenous
WQI indirect
INF individual
BLT infantry
ECJ infiltrate-d-r-s
WPE infiltrate-g-on
JMD inflict
NDU inform-d-ation
PWT infrared

RYX June
LPI jury
BNC just
GAK k
JFL keep-s-ing
BGQ kerosene
AYQ key
IER kidney
CUD kill
WIO killed in act
LMB kilometer-s
LJG kit
VTA knife
LXJ knot
VIF know-n-ing
ODR knowledge
EJE l
WAB labor-ed-ing
CDM lacerate
XOA lamp
RIT land-ed-ing
SUD large

ULN love
FGJ low-er-est
OLM lower side band
YFC lunch
SWI ly
TRI m
XJA machinegun-s
RTR made
KIM magnetic
EYW magnify-ing
ENF main
GTO main body
CIH main supply rt
YVS maintain-ed
DHG maintain-ing-s
MMU maintenance
LOB make-g/made
DIF male
VOE man-ned-ning
CNL maneuver-ng-s
IYE manual
VSV many

Ops-Code TEXT TO CODE Serial No. 001 Page 5

FLX map-ped-ping-s
HPN March/march
XXK market
NTW MARS
SUB match-d-s-ing
MSA max(range)
LUB May
SBL me
MOI meal
WBL means
RQJ measure-d
BQJ measure-ment
EFN meat
BIU mechanic-al
UIW mechanize-d
BKE medical
GKS medium
GBY medivac needed
AST meet (me)(at)
BXG meet-ing
EHH message-s

DOX MRE
HCH multiple
TYE music
BNX my
MKS my location is
OXR n
CMU nail
MSS narrow
VHG near
JAV need-ed-ing
OXM net-ted-ting
CDS network
JMI neutral-ize
EPN neutralize-d-ng
VLA new
PVK news
SPX newspaper
BXW night-s
CXV night-vision
NUW nine
ETM no fire line

TCI or
MUB orange
VLW order-ed-ing-s
CBI ordnance
TCH organize-ation
UWC organize-d
VXC orient-g-ation
CWW origin-al-ate
FSX other
PCC our
REA out(of)
TKS outpost
XHR over-age
KWH over-run
OEO overlay
MTO own
HKG p
XYN pacific
TGK packet
XGM PACTOR
KQW panel-s

VHK messenger
MUG meter-s
QTD MFSK
COF microphone
DDL might
KKY mile-s
TCR milk
XJG mine-d(field)
MRB minim-ize-d
NRH minimum
CBT minor
FOD minute-s
RAB miss-ed-ing
XUI missile-s
JGD missing in act
SOM mission-s
KQI mobile-ity
EWN modem
NYV modulation
QWX Monday
ANT money
TRC moon
HLF morale
MVR more
EUX more
KTS morning
QGO Morse
MCX most
CBK mother
NIC motor-ized
YCT mount-ed-ing
DKT mountain-ous
DHO move-d-ing
XSR move-ment

LOG no/non/neg
BSU none
AXL north-ern(of)
NKG not
WFQ not later than
PVI note
KOW notebook
SCH nothing
QHO notify-ied-ies
IQD November
HRY now
UFJ nuclear
CPK number-s
QDA o
MSG objective-s
QUR observation(post)
SDF observe-d-r
TYT observe-s-ing
CQJ obstacle-s
MFW occupy-ation
GYD occupy-ed-ing
VUP October
VCD odor
IEK of
PNS offense-ive
DEV officer-s
ALS oil
WJX on
SRA one
JJQ only
NCL open-ed-ing
AEC operate-ion-s
UHV operate-s-al
OYD opportunity-s

XDN pants
OAM paratroop-s
MOQ part-s
SWP party-ies
LUH pass-able
ABT pass-age-ing
SSC pass-es-ed
MFJ password
QSG past
UDV patrol-ed-ing-s
XVC PBBS
YEM pen
BHN pencile
MQV penetrate-d-ing
BDL penetrate-ion
TIH people
TQJ pepper
SDN per
ADM percent-age
HBU perimeter
QOF period
TNU permission
VOM permit-ed-ing
XIA PGP
OVB phase
DES photograph-ed-er-ing
LES pick
NLI pick up
MGD piece-d-ing-s
QMC pipe line
VCY pistol
YAR place-d-ing
RCT plan-s-ed-ing
ARR platoons

Ops-Code TEXT TO CODE   Serial No. 001   Page 6

NWR point-s-ed-ing
FCA police
RLR port(harbor)
LKP position-s-ed
GVU postpone-d-s
OMW postpone-ing
LSR possible
LFK post office
DJL post-s

JGS railroad-s
QDS rain
MKW range
EBV rate
NME ration-s-ed
DSP reach-ed-s-ing
VQO read
HWQ ready
FAC rear

EGP ridge-d-s
EDV rifle-s
UVG right(of)
MGJ riot-s-ed
FSW river-stream-s
RWE road-s
AKG roadblock
DNP round-s
VRY route-s-d-ing

IGE pound-s
EHO powder
KOH preparation
FJA prepare-d-ing-s
UFX prevent
QVH priest
BRW primary
ERP print
CUI printer
NPJ priority
XBJ privacy
VVW private
VXQ prize-s of war
DNO probable
EWS process-ed-ing
EMA professional
ANQ progress-ed-ing
SNL propane
RFH protect-d-s-ing
NIY protect-ed-ion
AVG protect-ive
VHA provide-d-ing
YOP provide-sion
BND province
JXK PSK
DEU psych
SJP pure
UTL purification
GWD pursue-d-ing-t
EDQ put
HYQ q
FUS quell
QDO quest
QFY question-ed
VVD quiet
EUL quit
GLD quiz
FBF r
LHJ rad(rad/hr)
RLJ radar
VVP radiation
DRG radiator
SJD radio-s
OOC radiological
FXN raid-er-ing-s
BLC railhead
PIQ receive-r-d-ng
HPH recognize-d-ing
DAI recon
YNE record
YCV recorder
JAS recover-ed-y
RKJ red
VXP reduce-g-tion
VJB reference (to)
BOE reference-d-s
AOD refuge
EEK refugee(s)
NJP regiment-ed-s
XTO registration
ODW regroup-ed-ng
GIG reinf-ed-s-ing
CTP reinf-ment-s
VQP relay-ed-ing
IDA release-d-ing-s
DUN relief/relieve
TDV remain-ing
ECC rendezvous
XIG reorganize-d-ation-s
VLJ repair-ed-ing
YWL repeat-ed-ing
HBR repeater (radio)
AWQ repel-ed-ing-s
CWK replace-d-ing
OBT replace-ment
RKE report-ed-ing
BCX report-s(to)
LHR repulse-d-ing
WOW request-d-g-s
CIQ require-d-ing
KDF require-ment
QLU rescue
JOH reserve-d-g-s
NDF resist-ed-ance
CBL resist-ing
WNF rest
AAU restore-s-d
FJB restrict-ion
DKL restrict-ive
XED result-s-ed-ng
FPD resupply-d-s-g
YJT return-ed-ing
SXM RTTY
VWO rucksack
UIT rug
BOX ruin-s-ed
CUQ s
BPO sad
FIR safe-r-ty
YFG said
RAD salt
UPW Saturday
WVN saw
DEO schedule-d-ing-s
GJL school
LCM scope
TCB scout-ed-ing-s
PDD screen-d-ing-s
HBJ search-d-ng-s
YFU second-s-ary
ODI secret
HLI section-s
NJS sector-s
DJS secure-d-ing
QAB secure-ity-s
ERF send-ing/sent
DAP separate-d-ion
EEN September
GIX sequence
LXG service-d-able
XSC seven
VBE shall
IGD shell-s-ing-ed
GAR shelter
DEX shirt
OBY shoot-ing/shot
AQP short-ed
LJX shortwave
UAV shoulder
AYO sick
OIA sieze-d-ing
ERS sieze-ure-s
BRN sight-ed-ing
XSD sign-ed-ing
NQF signal-ed-ing
OPJ silent-ly-ce
OCM simplex
UFL simulate-d-ing

Ops-Code TEXT TO CODE Serial No. 001 Page 7

DQX sister
FNG sit/sitrep
KUW site-s
VSE six
DGS sleep-ing
DSM slope-s-ing-d
WEN slow-ed-ing
VAI small-er-est
DQJ smoke-d-ing-s
TDX sniper
SAO snow
YMD so
LQV socks
QOU solar
BJO solder
DTX soldier
YXT some
ISG soon-er-est
DMT sorry
DBQ sortie-s
TRH south-ern(of)
UTV speak
VXR spearhead
GGV spice
WSP spine
JRF sporadic
PVJ spot-ted
NHG squad-s
THT stand-ing(by)
CXW start-ed-ing
SBJ station-ed-s
CNP status(report)
EBC stay-ed-ing
BRG stole-n
DPV stop-ped-ing
RDI storm
IKN storm-ing-(s)
RLY street
PGD strength-en
AEO strike/struck
IDR strong-er-est
GPO submit-ed-ing
FUE success-ful-ly
VQS such
DDP Sunday

BBJ survey-ed-s
UQJ survival
IEE suspect-d-ing
CCX swallow-ed(s)
OWN switch-ed-ing
UGL swollen
BWR symmetric
BSK t
AAJ tablet
EAH tactic-s-al
IJV take-ing/took
GBF talk
QVD tank destroyer-s
MTS tank-er-s
QQY tape
BWP target-s
PWX task(force)
FEJ taxi
VHX team-s-ed
KBB tear-gas
DCG ted
QBF teen
VBN teeth
QPH telephone-s
UQS television
KOF temple
ICB tentative
IWX terminate-d-g
NNQ terrain
TGR territory
ALU terrorism
YIF terrorists
WJN that
WEU the
NPM their
SKW them
FYT then
DTP there
WHI these
RST they
DVY this
XMX those
THP thousand
KUN three
XXF THROB

PIF tomorrow
UMX ton-s-age
CVO tone
GTP tonight
NBX tooth
BKC top secret
VAG total
GDX tow-ed-ing
EJY toward
VNQ town/village-s
ESL trace-r
JWD traffic
EJN trail-er-d-ing-s
EXU train-ed-ing-s
SSM track-ed-ing
SGE transmission
TFF transmit-ed
UTT transport-ed
FFN transport-g-s
TEA transportation
NPD trapped
NKS trouble
NNX truck-ed-s-ing
ILT Tuesday
EXV turn-ed-ing
PLG two
MOA u
NIP unable
VBP unarmed
TXA unauthorized
XHD under
NVB undercover
QPU understand
UNF understood
NAW unharmed
PPQ unidentified
MOP unite-d-s-ing
BGR until(further notice
SEJ unusual
LYS up
QXP upper side band
IPT us
YYT use-d-g
NQK UTC
OVQ V

XCX sunrise
NLY sunset
RPX superior-ity
VFI supplement-ed-ary
HQN supply-d-s-g
QGK support-s-d-ing
ERG surface-d-ing
KQF surgery
ITL surround-ed-ing
RMS surveillance
DMA through
BEG thunder
VPT Thursday
BQH tide
RTT time
PWA time-d-ing
CXL tion
XSI tire
SDA to(be)
NWC today
HUG vaccine
RMT valley
FES vegetable
CKT vehicle-s
MCJ verification
LKE verify-ied
MAH vertical
NQG very
VOX video
TRD violate-s-d

Ops-Code TEXT TO CODE Serial No. 001 Page 8

NMJ violet
XUL virus
FNO visible-ity
LIJ volt-s
ITA voltage
KBW vulnerable
GNK w
WFW wagon
FEU wait-ed-ing
HJE war
FIK warehouse
HHM was/were
YVH water
XGJ watts
NID wax
GOE waypoint
PMC we/us
UOD weak-en-ness
ANK weapon-s
HLQ weather
CYE Wednesday
EHY week-s
ARS weekend
OLG well
HBW west-ern(of)
EGV what
VVQ wheel-ed
HHW when
HRD where
IIC which
GYV white
TIX who
YYW wife
OSA yet
KMR you-r
GRV z
YNQ zap
WYO zero
ALX zip
WQX zipper
XLC zone-s-ing
AHG zoo

BRA wild
LGO wilderness
NKE will
ENM wind
RRN WinLink
JTF wire-s-d-ing
IQI with
OTU withdraw-ing-s
GEY wood-s-ed
EHF work-s-ed-ing
WOD wound-ed-s
RYA wreck-ed-s
NES wrist
WYU wrong
ADK x
XJC x-mit
BWJ x-ray
CFU y
MCB yard-s
WHR yellow
FPR yes
IMY yesterday

Ops-Code CODE TO TEXT   Serial No. 001   Page 9

AAJ tablet
AAR fruit
AAS as
AAU restore-s-d
ABO et
ABT pass-age-ing
ACB h
ACT conserve-d
ADD bite
ADK x
ADM percent-age
ADY April
AEC operate-ion-s
AEO strike/struck
AEX clear-ed-ance
AFB is
AFK abort-s-ed-ing
AFL c
AGD estimate-ion
AGO evasion
AHG zoo
AWQ repel-ed-ing-s
AXI impassable
AXK great
AXL north-ern(of)
AYA brother
AYH extend-extent
AYO sick
AYQ key
BBC library
BBF his
BBJ survey-ed-s
BCX report-s(to)
BCY fog
BDB battalion-s
BDI embrace-d-ing
BDL penetrate-ion
BDO interfere-nce
BEG thunder
BEJ ied
BFA interfere-d-ing
BFF airdrop-s
BQJ measure-ment
BQL fight-er-ing
BRA wild
BRC liaison
BRF handi-talkie
BRG stole-n
BRN sight-ed-ing
BRR identification
BRW primary
BSK t
BSU none
BSV bike
BTB break/broken
BTJ anti-
BUJ cancel-led-s
BUK dangerous
BVF boobytrap
BVN insect
BWC interval
BWD er
BWJ x-ray

AHW fade-ed-ing
AJQ altitude
AJS am in pos (to)
AKG roadblock
AKO collect-ed-ion
AKP far
ALS oil
ALU terrorism
ALX zip
AME barrage-s
ANG foe
ANK weapon-s
ANQ progress-ed-ing
ANT money
AOD refuge
APK arm-s-ed
APN cost
AQB interrogate-d
AQN IDY
AQP short-ed
AQX hour-s-ly
ARH block-ed-ing
ARR platoons
ARS weekend
ARV enlist-ed-ing
ASH confidential
ASL intersection
AST meet (me)(at)
ASX home
ATR battery-ies
AUQ delay-ed-ing-s
AVF accomplish-ed
AVG protect-ive
AVL border

BFH crane
BFR direct-d-on-ng
BGQ kerosene
BGR until(further notice
BHN pencil
BIR light-er-ng-d-s
BIU mechanic-al
BJO solder
BKC top secret
BKE medical
BKF approach-ed-s
BLC railhead
BLT infantry
BLY city
BMA clearing-s
BMO ice
BNC just
BND province
BNJ approve-al-d-ing
BNP bad
BNX my
BOC decontaminate-d-ion
BOE reference-d-s
BOM item-s
BOX ruin-s-ed
BPG headquarters
BPI final
BPJ escape-d-ing
BPO sad
BPP ambulance
BPS aircraft
BPV detach-ment-s
BPX ask
BQH tide

BWP target-s
BWR symmetric
BXG meet-ing
BXW night-s
CAU designate-ion
CBI ordnance
CBK mother
CBL resist-ing
CBN handbook
CBR car
CBT minor
CCD administrative
CCI assist-ance
CCX swallow-ed(s)
CCY attach-ed-ing
CDA by
CDL crossroad-s
CDM lacerate
CDR capture-d-ing
CDS network
CEQ able/ability
CEV decrease-d-ing
CFQ alley
CFU y
CGI blister
CHQ leaflet-s
CIC limit-ed-ing-s
CIH main supply rt
CIL arrive-al-d-ing
CIQ require-d-ing
CJA atlantic
CJJ i
CJL company
CKN expect-ation-s

Ops-Code CODE TO TEXT   Serial No. 001   Page 10

CKT vehicle-s
CMG cup
CMU nail
CND gate
CNK drive
CNL maneuver-ng-s
CNP status(report)
COF microphone
CPC commit-ment

DJL post-s
DJS secure-d-ing
DKL restrict-ive
DKT mountain-ous
DKU fail-ed-ing
DLE like-s
DMA through
DMR aviator/pilot
DMT sorry

EIL check-d-ing-s
EIY hole
EJE l
EJN trail-er-d-ing-s
EJY toward
EKQ local-ize-d
EKY computer
EMA professional
EMI adjacent

CPK number-s
CPL eight
CQJ obstacle-s
CQQ depot
CSH administration
CTP reinf-ment-s
CTW counterattack-ed-s
CTX connect-ion
CUA extend-d-g-ive
CUD kill
CUI printer
CUJ hold-er-ing
CUQ s
CUW dispose-d-al
CVO tone
CVS exploit-ed-ing
CVV deadline-d-ing
CWC government
CWK replace-d-ing
CWW origin-al-ate
CXL tion
CXV night-vision
CXW start-ed-ing
CYE Wednesday
DAI recon
DAP separate-d-ion
DAW burst-ing-s
DBG ground-s-ed
DBO add-ing-s
DBP enemy
DBQ sortie-s
DCG ted
DDA lock
DDL might
DDP Sunday
DEO schedule-d-ing-s
DES photograph-ed-er-ing
DEU psych
DEV officer-s
DEX shirt
DGP headphones
DGS sleep-ing
DHG maintain-ing-s
DHO move-d-ing
DIF male
DJE b
DNO probable
DNP round-s
DOX MRE
DPV stop-ped-ing
DPX ies
DQJ smoke-d-ing-s
DQR join-ed-ing
DQX sister
DRG radiator
DRR deploy-ed-ing
DRS load-s-ed
DSM slope-s-ing-d
DSP reach-ed-s-ing
DTJ fracture
DTP there
DTS leave/left
DTX soldier
DUA freeze (ing)
DUJ FEMA
DUM authorization
DUN relief/relieve
DVX convoy-s
DVY this
DYG him
EAH tactic-s-al
EBC stay-ed-ing
EBV rate
ECB law
ECC rendezvous
ECJ infiltrate-d-r-s
EDC ditch
EDQ put
EDT boots
EDV rifle-s
EDW complete-d-ing
EEK refugee(s)
EEN September
EFN meat
EGB hundred
EGP ridge-d-s
EGV what
EHF work-s-ed-ing
EHH message-s
EHO powder
EHY week-s
EII combat
ENF main
ENM wind
ENO connect-ed-ing
ENU alternate
EOF an
EOM jeep
EPN neutralize-d-ng
EQD letter
EQG corporation
EQX litter-s
ERF send-ing/sent
ERG surface-d-ing
ERM frequency-ies
ERP print
ERS sieze-ure-s
ESL trace-r
ESY group
ETL FRS
ETM no fire line
EUL quit
EUX more
EVE emergency
EWN modem
EWS procees-ed-ing
EXU train-ed-ing-s
EXV turn-ed-ing
EYW magnify-ing
FAC rear
FAU good
FBF r
FBN barometer
FBP drug-s
FCA police
FDV boundary-ies
FEJ taxi
FEM internal
FEQ encounter-d-ing
FES vegetable
FEU wait-ed-ing
FFI for
FFN transport-g-s
FGJ low-er-est
FHG embank-ment
FIK warehouse
FIR safe-r-ty
FJA prepare-d-ing-s

Ops-Code CODE TO TEXT Serial No. 001 Page 11

FJB restrict-ion
FKG inadequate
FLB change to/from ciphe
FLH isolate-d-ing
FLX map-ped-ping-s
FMX issue-d-ing-s
FNG sit/sitrep
FNO visible-ity
FNU fire-d-ing-s
FOD minute-s
FPD resupply-d-s-g
FPN church
FPR yes
FRE bulldozer
FRS bit
FSW river-stream-s
FSX other
FUE success-ful-ly
FUN immediately
FUS quell
FWJ conservation
FWK intact
FXN raid-er-ing-s
FYT then
GAI hostile
GAK k
GAR shelter
GBF talk
GBT deliver-ed-ing
GBY medivac needed
GDM chance
GDX tow-ed-ing
GEY wood-s-ed
GGV spice
GGW disregard-d-ing
GHD blue
GHK craft
GHY explosive-s
GIG reinf-ed-s-ing
GIX sequence
GJL school
GKS medium
GLD quiz
GLG gunpowder
GLQ grade-d-r-ing
GQK compound
GQM hill/mt/rise-s
GQN culvert
GRV z
GTO main body
GTP tonight
GVA admit
GVU pospone-d-s
GWA come-ing
GWD pursue-d-ing-t
GXU between
GYD occupy-ed-ing
GYU list
GYV white
HBF hear-ed
HBJ search-d-ng-s
HBR repeater (radio)
HBU perimeter
HBW west-ern(of)
HCH multiple
HDF effect-ive-s
HFM en
HHC bivouac-ed-ing
HHM was/were
HHW when
HHX counterattack-ing
HJE war
HKG p
HLF morale
HLI section-s
HLL it
HLQ weather
HLV agent-s
HME at (once)
HPH recognize-d-ing
HPN March/march
HQN supply-d-s-g
HRD where
HRY now
HSN arrange-ment
HTI lead-er-ing
HTP amputate
HUG vaccine
HVY insert-s-ed-ing
HWQ ready
IFU dig/dug (in)
IFW bed
IGD shell-s-ing-ed
IGE pound-s
IGQ fuse-d
IIB engine
IIC which
IJV take-ing/took
IJW defense-ive
IKN storm-ing-s
ILR expect-ed-ing
ILT Tuesday
IMY yesterday
INF individual
ING increase-d-g-s
IPF capacity-ies
IPT us
IQB least
IQD November
IQI with
IQN attach-ment
ISG soon-er-est
ISL distance-s
ITA voltage
ITD continue-d-ing
ITL surround-ed-ing
IWG cut-off
IWX terminate-d-g
IXN critical-ly
IYE manual
JAS recover-ed-y
JAV need-ed-ing
JBI airborne DF platform
JBY favor-s-ed
JCY hidden
JFL keep-s-ing
JGD missing in act
JGS railroad-s
JGX here
JIA capsule
JJK chemical-s
JJQ only
JKC August
JKE CB radio
JKO comma

GML five
GMS friend-ly
GNK w
GNL forward-ed-ing
GNY locate-ion(of)
GOE waypoint
GOU arrow
GPF back-ed-ing
GPO submit-ed-ing
GPV cease-s-d-ing
HYQ q
IAE bomb-d-r-ng-s
ICB tentative
ICX gasoline
IDA release-d-ing-s
IDR strong-er-est
IEE suspect-d-ing
IEK of
IER kidney
IEU agree
JKT defend-ed-g-s
JLY class-ify
JMD inflict
JMI neutral-ize
JML asap
JNU camouflage-d
JOE high speed burst
JOH reserve-d-g-s
JRF sporadic
JTA concentrate-s-ion

JTF wire-s-d-ing
JVM before
JWD traffic
JWS all
JXK PSK
JYE f
JYM high-er-est
JYP had
KBB tear-gas
KBW vulnerable
KCB fortify-ied
KCQ elevate-d-ion
KCU command-er-ed-ing-s
KDF require-ment
KDQ destination
KEM collapse
KFE advance-d-ing
KGD alt supply pt
KGE junction-s(road)
KGH harass-ment
KGR band
KHV call
KIE drip
KIG highway-s
KIM magnetic
KKG Friday
KKN lay/laid
KKO cook
KKY mile-s
KMB damage-d-ing
KMR you-r
KOF temple
KOH preparation
LDV engineer-d-ing-s
LEP detach-ed-es
LES pick
LET improve-ment
LFC elk
LFJ general-ize
LFK post office
LGA frag
LGO wilderness
LHJ rad(rad/hr)
LHR repulse-d-ing
LID assault
LIH a
LII ammunition
LIJ volt-s
LJG kit
LJK district-s
LJX shortwave
LKE verify-ied
LKP position-s-ed
LLD in-accordance-with
LMB kilometer-s
LNA coast-s-al
LNJ and
LNN could
LOB make-g/made
LOG no/non/neg
LPI jury
LQV socks
LRB electric-al-ity
LRI helicopter-s
LSR possible
LSY abdomen
MGD piece-d-ing-s
MGJ riot-s-ed
MGO he
MGS interrogate-ion
MHN beer
MHQ fuel
MIQ incident-al-s
MKS my location is
MKV go-ing(to)
MKW range
MMS inch-es
MMU maintenance
MNM future
MOA u
MOI meal
MOP unit-ed-s-ing
MOQ part-s
MOS fold-ing
MQE clean-ed-ing
MQV penetrate-d-ing
MRB minim-ize-d
MRS force-s-d-ing
MSA max(range)
MSG objective-s
MSS narrow
MTN determine
MTO own
MTS tank-er-s
MTV from
MUB orange
MUF left(of)
MUG meter-s
MUU listen-ed-ing

KOO consolidate-d
KOU look(for)
KOW notebook
KPM CW
KPV harass-ed-ing
KQF surgery
KQI mobile-ity
KQU GPS
KQW panel-s
KSJ feet/foot
KSL intend-ed-tion
KTD caliber
KTS morning
KUN three
KUW site-s
KWH over-run
KYH accomplishing
LAD divert-ed-ion
LAG communicate-d-ions-s
LAO deer
LCM scope
LDH boy

LUB May
LUH pass-able
LWB e-mail
LWF approach-ing
LWV indigenous
LWY cross-ed-ing
LXG service-d-able
LXJ knot
LYS up
MAH vertical
MAV authorize-d-ng
MAY acknowledge-d-ment
MCB yard-s
MCJ verification
MCX most
MDA cache'
MDT canal-s
MDV DTMF
MEQ contact-ing
MFJ password
MFL deception
MFW occupy-ation

MVR more
MVY hate
MXI beach-ed-s-ing
MYY east-ern (of)
NAW unharmed
NBT log
NBX tooth
NCF downed acft
NCL open-ed-ing
NCM continue-ation
NDF resist-ed-ance
NDM equip-ing-s
NDR if
NDU inform-d-ation
NES wrist
NFC dog
NFY improve-d-ing
NGB in
NHG squad-s
NHI control-d-ing-s
NHT commit-ted-ing
NIC motor-ized

Ops-Code CODE TO TEXT Serial No. 001 Page 13

NID wax
NIP unable
NIY protect-ed-ion
NJL immobile-ize-d-ing
NJP regiment-ed-s
NJS sector-s
NKD contaminate-ion-ing
NKE will
NKG not
NKS trouble
NKX column-s
NLF baud
NLI pick up
NLJ cloth-es-ing
NLN father
NLY sunset
NME ration-s-ed
NMJ violet
NML gas-ing-ed
NMO jump-ed-g(off)
NNQ terrain

OFX contaminate-d
OGG impact
OHB coord-s-d-ing
OIA sieze-d-ing
OJQ July
OLG well
OLM lower side band
OLV direction finding
OMU facility
OMW pospone-ing
ONA accident-al-s
ONT diesel
ONU January
ONV forest-s
OOC radiological
OOT glass
OPC level
OPJ silent-ly-ce
OPK bridge-d-ing-s
OSA yet
OSM four

PNS offense-ive
PNX improvised
PPQ unidentified
PQA article
PQJ disperse-d-g-l
PRA envelope
PSA add-ed-itional
PTK breast
PUV leg
PVC act-ive-ivity-s
PVF guard-ed-ing-s
PVI note
PVJ spot-ted
PVK news
PWA time-d-ing
PWE employ-ed-ing
PWR impossible
PWT infrared
PWX task(force)
PXF death
PXW GMRS

NNX truck-ed-s-ing
NOH destroy-d-ing-s
NON child/children
NPD trapped
NPJ priority
NPM their
NQF signal-ed-ing
NQG very
NQK UTC
NQM flesh
NRH minimum
NRY d
NSG heavy
NSM airmobile
NTW MARS
NUR automatic
NUW nine
NVB undercover
NWC today
NWR point-s-ed-ing
NYL during
NYT extract-s-d-ing
NYV modulation
OAM paratroop-s
OBT replace-ment
OBY shoot-ing/shot
OCM simplex
ODI secret
ODR knowledge
ODW regroup-ed-ng
OEO overlay
OEU frequency modulation
OFP carry-ied-iers
OFT female
OTF extend-sion
OTL blow-ing (up)
OTU withdraw-ing-s
OVB phase
OVQ V
OWI book
OWN switch-ed-ing
OXM net-ted-ting
OXR n
OYD opportunity-s
OYT flank-ed-ing-s
OYY crystal
PAO field-s
PBO hammer
PCC our
PCQ alert-ed-ing
PDD screen-d-ing-s
PDO girl
PDU breakfast
PDV encrypt-ed-tion
PGB army-ies
PGD strength-en
PHF depart-d-g-ure
PIF tomorrow
PIN expose-d-ing
PIQ receive-r-d-ng
PJK dinner
PKE contact-ed-s
PKI exercise
PKN complete-ion
PLG two
PMC we/us
PMT approximate-ly
PMV get-ting
PYE charge-d-s
PYN ed
QAB secure-ity-s
QAH ache
QAO es
QBF teen
QDA o
QDO quest
QDS rain
QEP area-s
QEY end-ed-ing
QFY question-ed
QGK support-s-d-ing
QGO Morse
QGU grid(coord)
QHB antenna
QHO notify-ied-ies
QKQ deny-ied-ies
QKT flare-d-s
QLP line-d-s(of)
QLU rescue
QMC pipe line
QMD at
QMS channel
QNN attack-ed-ng-s
QNS available
QNT arrange-d-ing
QOF period
QOU solar
QOW disable-d-ing
QPH telephone-s
QPU understand
QPW eliminate-d-ing-s
QQY tape

Ops-Code CODE TO TEXT   Serial No. 001   Page 14

QRC gallon-s
QSF divide
QSG past
QTB December
QTD MFSK
QTF deploy-ment-s
QTQ checkpoint
QUR observation(post)
QVD tank destroyer-s
RUS age
RVE have/has
RVX day-s/date-s
RWA battery charger
RWE road-s
RWH challenge-d-ing
RWI binoculars
RXG her
RYA wreck-ed-s
TBO axe
TBX install-d-ation
TCB scout-ed-ing-s
TCH organize-ation
TCI or
TCR milk
TDV remain-ing
TDX sniper
TEA transportation

QVH priest
QWX Monday
QXH concentrate-d-ing
QXP upper side band
QXV hook
QYA candle
RAB miss-ed-ing
RAD salt
RBP conduct-ed-ing
RBR air (fld/port)
RBT bug(s)
RCL do
RCT plan-s-ed-ing
RDI storm
REA out(of)
RFH protect-d-s-ing
RFL held
RGL apply
RGQ exhaust-ed-ing
RIF flexible-ity
RIL bone
RIT land-ed-ing
RKE report-ed-ing
RKF favor-able-ing
RKJ red
RLI radar
RLR port(harbor)
RLY street
RMS surveillance
RMT valley
RNI desk
RNT flame-d-ing-s
RPU begin
RPX superior-ity
RQJ measure-d
RQK found
RQM jacket
RQP change
RRN WinLink
RST they
RSU (SPACE / NULL)
RTB guide-ance
RTP civilian-s
RTR made
RTT time
RUN down-ed
RYE jam-med-ing
RYX June
SAO snow
SAP capable-itiy
SBE disapprove-d-al
SBJ station-ed-s
SBL me
SBO belt
SCH nothing
SDA to(be)
SDF observe-d-r
SDN per
SDR hospital-s
SEA life
SEF front-age-ally
SEJ unusual
SGE transmission
SIX blow-n
SJD radio-s
SJP pure
SKW them
SMH job
SMN displace-d-ing
SNL propane
SNN intercept-d-or
SOD hand
SOF envelop-ed-ing
SOM mission-s
SPG FeldHell
SPR demonstrate-d-s-ion
SPX newspaper
SQG assist-ed-ing
SRA one
SSC pass-es-ed
SSM trak-ed-ing
SUB match-d-s-ing
SUD large
SVF logistic-al-s
SVI fallout
SVW division-s
SWI ly
SWP party-ies
SXC j
SXM RTTY
TAB collect-ing-s
TAT close-d-ing
TFB instrumental
TFF transmit-ed
TGK packet
TGR territory
THP thousand
THT stand-ing(by)
THX bow
TIH people
TIX who
TKS outpost
TKW classification
TLH box
TMF locate-d-ing-s
TNU permission
TOF CTCSS
TON battle-s
TQJ pepper
TRC moon
TRD violate-s-d
TRH south-ern(of)
TRI m
TRM alarm
TSG eta your loc
TTM consolidation
TUQ base-d-s-ing
TXA unauthorized
TYE music
TYT observe-s-ing
TYV execute-d-ing
UAR assemble-d-ing
UAV shoulder
UBL barrier
UBP launch-d-r-ing
UCB emplace-ment
UCK below
UCV camera
UDV patrol-ed-ing-s
UFJ nuclear
UFL simulate-d-ing
UFW conversation
UFX prevent
UGC corps
UGJ drill
UGL swollen
UGQ imminent
UHG extra

Ops-Code CODE TO TEXT Serial No. 001 Page 15

UHM cover-ed-ing
UHV operate-s-al
UIT rug
UIW mechanize-d
UJJ intercept-ion-s
UKM fish
UKR azimuth (of)
ULN love
UMF bleed
UMX ton-s-age
UNF understood
UNU circle
UOD weak-en-ness
UPQ fatal
UPW Saturday
UQJ survival
UQS television
UTF flight
UTL purification
UTT transport-ed
UTV speak
UUL element-s
UUT initial-ed-ly
UVC bunker-s
UVG right(of)
UVJ ambush-ed-ing
UVO estimate-d-ing
UVW expedite-d-ing
UWC organize-d
UWT hit-ting
UWU correct-ed-ing
UXE ing
VAG total
VAI small-er-est
VAQ lateral
VBE shall
VBN teeth
VBP unarmed
VBV gun-s
VCD odor
VCY pistol
VDJ darkness
VDW can
VEA barbed wire
VEQ amateur
VIF know-n-ing
VJB reference (to)
VJI green
VJU alive
VKC compromise-d-ing
VKL danger
VLA new
VLJ repair-ed-ing
VLU alternate commo
VLW order-ed-ing-s
VNB be-en-ing
VNQ town/village-s
VNS bank
VOE man-ned-ning
VOM permit-ed-ing
VOX video
VPN cave
VPT Thursday
VQO read
VQP relay-ed-ing
VQS such
VRU last
VRY route-s-d-ing
VSD estab-ed contact at(
VSE six
VSP identify-ed
VSV many
VTA knife
VTN ball
VTO forearm
VUP October
VVB how
VVD quiet
VVI amplitude modulation
VVP radiation
VVQ wheel-ed
VVW private
VWE family
VWO rucksack
VXC orient-g-ation
VXN explode-d-s
VXP reduce-g-tion
VXQ prize-s of war
VXR spearhead
VXT axis (of)
WGE error
WHI these
WHR yellow
WIO killed in act
WJC ill-ness
WJN that
WJX on
WLY consolidate-ng
WMF halt-ed
WNC above
WNF rest
WNW displace-ment
WOD wound-ed-s
WOM guide-d-s-ing
WOW request-d-g-s
WPE infiltrate-g-on
WQI indirect
WQX zipper
WRM ALE
WRT but
WSP spine
WSY ford-able
WVE been
WVH degree-s
WVK delete-d-ion
WVN saw
WWU fortify-ication
WXW g
WYC deputy-ies
WYO zero
WYU wrong
WYV generator(set)
WYY about
XAD designate-d
XBG cable-s
XBJ privacy
XCX sunrise
XDN pants
XED result-s-ed-ng
XFY are/is
XGJ watts
XGM PACTOR
XHD under
XHM hide
XHR over-age

VEV February
VEY joint
VFI supplement-ed-ary
VGV condition-d-ing
VHA provide-d-ing
VHG near
VHK messenger
VHM dump-ed-ing
VHP entrench-ed
VHX team-s-ed

WAB labor-ed-ing
WBL means
WDK etd
WDX drop-ping-ped
WEN slow-ed-ing
WEU the
WFC clarification(of)
WFQ not later than
WFT answer-d-ing-s
WFW wagon

XIA PGP
XIG reorganize-d-ation-s
XIV food
XJA machinegun-s
XJC x-mit
XJG mine-d(field)
XKJ biological
XKR give-n-ing
XLC zone-s-ing
XME building-s

Ops-Code CODE TO TEXT Serial No. 001 Page 16

XMX those
XOA lamp
XPI Ham/ham
XQQ follow-ed-ing
XSC seven
XSD sign-ed-ing
XSI tire
XSR move-ment
XSX due
XTG am
XTO registration
XUF after
XUI missile-s
XUL virus
XVC PBBS
XVE e
XVI fled
XVY around
XWF act-ion-ing-s
XXF THROB
XXK market
XYN pacific
XYV camouflage-ing
XYY crushed
YAR place-d-ing
YBS camp
YCN execute-ion-s
YCT mount-ed-ing
YCV recorder
YDL boat-s
YEB evac-d-ing-ion
YED establish-d-ng
YEL loudspeaker

YVS maintain-ed
YWL repeat-ed-ing
YWR freedom
YXG coffee
YXT some
YYL automobile
YYT use-d-g
YYW wife
YYY beyond

YEM pen
YFC lunch
YFG said
YFU second-s-ary
YGA correct (me)
YIC demolition
YIF terrorists
YII loss-es-lost
YJM bypass-d-ing-s
YJT return-ed-ing
YLI counterpart
YMD so
YMQ demonstrate-d-ing
YNE record
YNQ zap
YOP provide-sion
YOR casualty-ies
YQC effect-ed-ing
YRX black
YSV head-ed-ing
YVH water
YVP driveway

APPENDIX III

Amateur Radio Technician Exam Question Pool

The following material was reprinted with permission from the AARL Web site.

* To obtain a copy of the graphics references that are to be used with this question pool, see the downloadable PDF graphic on this web page, or send a business sized SASE to the ARRL/VEC, 225 Main St, Newington CT 06111. Request the "2003 Technician class Question Pool Graphics". For $1.50 the ARRL VEC will supply a hardcopy of this pool and graphics without receiving an SASE.

* The questions contained within this pool must be used in all Technician examinations beginning July 1, 2003, and is intended to be used up through June 30, 2007.

* The correct answer position A,B,C,D appears in parenthesis following each question number [e.g., in T1A01 (B), position B contains the correct answer text].

Question Pool ELEMENT 2 - TECHNICIAN CLASS as released by Question Pool Committee National Conference of Volunteer Examiner Coordinators December 4, 2003

SUBELEMENT T1 - FCC Rules [5 Exam Questions — 5 Groups]

T1A Definition and purpose of Amateur Radio Service, Amateur-Satellite Service in places where the FCC regulates these services and elsewhere; Part 97 and FCC regulation of the amateur services; Penalties for unlicensed operation and for violating FCC rules; Prohibited transmissions.

T1A01 (B) [97]
Who makes and enforces the rules for the amateur service in the United States?
A. The Congress of the United States
B. The Federal Communications Commission (FCC)
C. The Volunteer Examiner Coordinators (VECs)
D. The Federal Bureau of Investigation (FBI)

T1A02 (D) [97.1]
What are two of the five fundamental purposes for the amateur service in the United States?
A. To protect historical radio data, and help the public understand radio history
B. To help foreign countries improve communication and technical skills, and encourage visits from foreign hams
C. To modernize radio schematic drawings, and increase the pool of electrical drafting people
D. To increase the number of trained radio operators and electronics experts, and improve international goodwill

T1A03 (D) [97.3a5]
What is the definition of an amateur station?
A. A radio station in a public radio service used for radiocommunications
B. A radio station using radiocommunications for a commercial purpose
C. A radio station using equipment for training new broadcast operators and technicians
D. A radio station in the amateur service used for radiocommunications

T1A04 (A) [97.113b]
When is an amateur station authorized to transmit information to the general public?
A. Never
B. Only when the operator is being paid
C. Only when the broadcast transmission lasts less than 1 hour
D. Only when the broadcast transmission lasts longer than 15 minutes

T1A05 (A) [97.113a4, 97.113e]
When is an amateur station authorized to transmit music?
A. Amateurs may not transmit music, except as an incidental part of an authorized rebroadcast of space shuttle communications
B. Only when the music produces no spurious emissions
C. Only when the music is used to jam an illegal transmission
D. Only when the music is above 1280 MHz, and the music is a live performance

T1A06 (C) [97.113a4, 97.211b, 97.217]
When is the transmission of codes or ciphers allowed to hide the meaning of a message transmitted by an amateur station?
A. Only during contests
B. Only during nationally declared emergencies
C. Codes and ciphers may not be used to obscure the meaning of a message, although there are special exceptions
D. Only when frequencies above 1280 MHz are used

T1A07 (B) [97.3a10, 97.113b]
Which of the following one-way communications may NOT be transmitted in the amateur service?
A. Telecommand to model craft
B. Broadcasts intended for reception by the general public
C. Brief transmissions to make adjustments to the station
D. Morse code practice

T1A08 (C) [97.3a40]
What is an amateur space station?
A. An amateur station operated on an unused frequency
B. An amateur station awaiting its new call letters from the FCC
C. An amateur station located more than 50 kilometers above the Earth's surface
D. An amateur station that communicates with the International Space Station

T1A09 (B) [97.207a]
Who may be the control operator of an amateur space station?
A. An amateur holding an Amateur Extra class operator license grant
B. Any licensed amateur operator
C. Anyone designated by the commander of the spacecraft
D. No one unless specifically authorized by the government

T1A10 (A) [97.113a4]
When may false or deceptive signals or communications be transmitted by an amateur station?

A. Never
B. When operating a beacon transmitter in a "fox hunt" exercise
C. When playing a harmless "practical joke"
D. When you need to hide the meaning of a message for secrecy

T1A11 (C) [97.119a]
When may an amateur station transmit unidentified communications?
A. Only during brief tests not meant as messages
B. Only when they do not interfere with others
C. Only when sent from a space station or to control a model craft
D. Only during two-way or third-party communications

T1A12 (A) [97.119a]
What is an amateur communication called that does NOT have the required station identification?
A. Unidentified communications or signals
B. Reluctance modulation
C. Test emission
D. Tactical communication

T1A13 (B) [97.3a23]
What is a transmission called that disturbs other communications?
A. Interrupted CW
B. Harmful interference
C. Transponder signals
D. Unidentified transmissions

T1A14 (A) [97.3a10]
What does the term broadcasting mean?
A. Transmissions intended for reception by the general public, either direct or relayed
B. Retransmission by automatic means of programs or signals from non-amateur stations
C. One-way radio communications, regardless of purpose or content
D. One-way or two-way radio communications between two or more stations

T1A15 (D) [97.113a4]
Why is indecent and obscene language prohibited in the Amateur Service?
A. Because it is offensive to some individuals
B. Because young children may intercept amateur communications with readily available receiving equipment
C. Because such language is specifically prohibited by FCC Rules
D. All of these choices are correct

T1A16 (B) [97.113a3]
Which of the following is a prohibited Amateur Radio transmission?
A. Using an autopatch to seek emergency assistance
B. Using an autopatch to pick up business messages
C. Using an autopatch to call for a tow truck
D. Using an autopatch to call home to say you are running late

T1B International aspect of Amateur Radio; International and domestic spectrum allocation; Spectrum sharing; International communications; reciprocal operation; International and domestic spectrum allocation; Spectrum sharing; International communications; reciprocal operation.

T1B01 (B) [97.301a]
What are the frequency limits of the 6-meter band in ITU Region 2?
A. 52.0 - 54.5 MHz
B. 50.0 - 54.0 MHz
C. 50.1 - 52.1 MHz
D. 50.0 - 56.0 MHz

T1B02 (A) [97.301a]
What are the frequency limits of the 2-meter band in ITU Region 2?
A. 144.0 - 148.0 MHz
B. 145.0 - 149.5 MHz
C. 144.1 - 146.5 MHz
D. 144.0 - 146.0 MHz

T1B03 (B) [97.301f]
What are the frequency limits of the 1.25-meter band in ITU Region 2?
A. 225.0 - 230.5 MHz

B. 222.0 - 225.0 MHz
C. 224.1 - 225.1 MHz
D. 220.0 - 226.0 MHz

T1B04 (C) [97.301a]
What are the frequency limits of the 70-centimeter band in ITU Region 2?
A. 430.0 - 440.0 MHz
B. 430.0 - 450.0 MHz
C. 420.0 - 450.0 MHz
D. 432.0 - 435.0 MHz

T1B05 (D) [97.301a]
What are the frequency limits of the 33-centimeter band in ITU Region 2?
A. 903 - 927 MHz
B. 905 - 925 MHz
C. 900 - 930 MHz
D. 902 - 928 MHz

T1B06 (B) [97.301a]
What are the frequency limits of the 23-centimeter band in ITU Region 2?
A. 1260 - 1270 MHz
B. 1240 - 1300 MHz
C. 1270 - 1295 MHz
D. 1240 - 1246 MHz

T1B07 (A) [97.301a]
What are the frequency limits of the 13-centimeter band in ITU Region 2?
A. 2300 - 2310 MHz and 2390 - 2450 MHz
B. 2300 - 2350 MHz and 2400 - 2450 MHz
C. 2350 - 2380 MHz and 2390 - 2450 MHz
D. 2300 - 2350 MHz and 2380 - 2450 MHz

T1B08 (C) [97.303]
If the FCC rules say that the amateur service is a secondary user of a frequency band, and another service is a primary user, what does this mean?
A. Nothing special; all users of a frequency band have equal rights to operate
B. Amateurs are only allowed to use the frequency band during emergencies
C. Amateurs are allowed to use the frequency band only if they do not cause harmful interference to primary users
D. Amateurs must increase transmitter power to overcome any interference caused by primary users

T1B09 (C) [97.101b]
What rule applies if two amateur stations want to use the same frequency?
A. The station operator with a lesser class of license must yield the frequency to a higher-class licensee
B. The station operator with a lower power output must yield the frequency to the station with a higher power output
C. Both station operators have an equal right to operate on the frequency
D. Station operators in ITU Regions 1 and 3 must yield the frequency to stations in ITU Region 2

T1B10 (D) [97.301e]
If you are operating on 28.400 MHz, in what amateur band are you operating?
A. 80 meters
B. 40 meters
C. 15 meters
D. 10 meters

T1B11 (D) [97.301f]
If you are operating on 223.50 MHz, in what amateur band are you operating?
A. 15 meters
B. 10 meters
C. 2 meters
D. 1.25 meters

T1B12 (D) [97.111a1]
When are you allowed to communicate with an amateur in a foreign country?
A. Only when the foreign amateur uses English
B. Only when you have permission from the FCC
C. Only when a third party agreement exists between the US and the foreign country
D. At any time, unless it is not allowed by either government

T1B13 (A) [97.303h]
If you are operating FM phone on the 23-cm

band and learn that you are interfering with a radiolocation station outside the US, what must you do?
A. Stop operating or take steps to eliminate this harmful interference
B. Nothing, because this band is allocated exclusively to the amateur service
C. Establish contact with the radiolocation station and ask them to change frequency
D. Change to CW mode, because this would not likely cause interference

T1B14 (A) [97.107]
What does it mean for an amateur station to operate under reciprocal operating authority?
A. The amateur is operating in a country other than his home country
B. The amateur is allowing a third party to talk to an amateur in another country
C. The amateur has permission to communicate in a foreign language
D. The amateur has permission to communicate with amateurs in another country

T1B15 (A) [97.301(f)(1)]
What are the frequency limits for the Amateur Radio service for stations located north of Line A in the 70-cm band?
A. 430 - 450 MHz
B. 420 - 450 MHz
C. 432 - 450 MHz
D. 440 - 450 MHz

T1C All about license grants; Station and operator license grant structure including responsibilities, basic differences; Privileges of the various operator license classes; License grant term; Modifying and renewing license grant; Grace period.

T1C01 (C) [97.5a]
Which of the following is required before you can operate an amateur station in the US?
A. You must hold an FCC operator's training permit for a licensed radio station
B. You must submit an FCC Form 605 together with a license examination fee
C. The FCC must grant you an amateur operator/primary station license
D. The FCC must issue you a Certificate of Successful Completion of Amateur Training

T1C02 (D) [97.9a]
What are the US amateur operator licenses that a new amateur might earn?
A. Novice, Technician, General, Advanced
B. Technician, Technician Plus, General, Advanced
C. Novice, Technician Plus, General, Advanced
D. Technician, Technician with Morse code, General, Amateur Extra

T1C03 (C) [97.5a]
How soon after you pass the examination elements required for your first Amateur Radio license may you transmit?
A. Immediately
B. 30 days after the test date
C. As soon as the FCC grants you a license and the data appears in the FCC's ULS data base
D. As soon as you receive your license from the FCC

T1C04 (A) [97.21a3i]
How soon before the expiration date of your license may you send the FCC a completed Form 605 or file with the Universal Licensing System on the World Wide Web for a renewal?
A. No more than 90 days
B. No more than 30 days
C. Within 6 to 9 months
D. Within 6 months to a year

T1C05 (C) [97.25a]
What is the normal term for an amateur station license grant?
A. 5 years
B. 7 years
C. 10 years
D. For the lifetime of the licensee

T1C06 (A) [97.21b]
What is the "grace period" during which the FCC will renew an expired 10-year license?
A. 2 years
B. 5 years

C. 10 years
D. There is no grace period

T1C07 (D) [97.103a]
What is your responsibility as a station licensee?
A. You must allow another amateur to operate your station upon request
B. You must be present whenever the station is operated
C. You must notify the FCC if another amateur acts as the control operator
D. You are responsible for the proper operation of the station in accordance with the FCC rules

T1C08 (B) [97.5d]
Where does a US amateur license allow you to operate?
A. Anywhere in the world
B. Wherever the amateur service is regulated by the FCC
C. Within 50 km of your primary station location
D. Only at the mailing address printed on your license

T1C09 (B) [97.113a3]
Under what conditions are amateur stations allowed to communicate with stations operating in other radio services?
A. Never; amateur stations are only permitted to communicate with other amateur stations
B. When authorized by the FCC or in an emergency
C. When communicating with stations in the Citizens Radio Service
D. When a commercial broadcast station is using Amateur Radio frequencies for newsgathering during a natural disaster

9
To what distance limit may Technician class licensees communicate?
A. Up to 200 miles
B. There is no distance limit
C. Only to line of sight contacts distances
D. Only to contacts inside the USA

T1C11 (A)
If you forget to renew your amateur license and it expires, may you continue to transmit?
A. No, transmitting is not allowed
B. Yes, but only if you identify using the suffix "GP"
C. Yes, but only during authorized nets
D. Yes, any time for up to two years (the "grace period" for renewal)

T1D Qualifying for a license; General eligibility; Purpose of examination; Examination elements; Upgrading operator license class; Element credit; Provision for physical disabilities.

T1D01 (A) [97.5b1]
Who can become an amateur licensee in the US?
A. Anyone except a representative of a foreign government
B. Only a citizen of the United States
C. Anyone except an employee of the US government
D. Anyone

T1D02 (D) [97.5b1]
What age must you be to hold an amateur license?
A. 14 years or older
B. 18 years or older
C. 70 years or younger
D. There are no age limits

T1D03 (D)
What government agency grants your Amateur Radio license?
A. The Department of Defense
B. The State Licensing Bureau
C. The Department of Commerce
D. The Federal Communications Commission

T1D04 (B) [97.501c]
What element credit is earned by passing the Technician class written examination?
A. Element 1
B. Element 2
C. Element 3
D. Element 4

T1D05 (C) [97.9b]
If you are a Technician licensee who has passed a Morse code exam, what is one document you can use to prove that you are authorized to use certain amateur frequencies below 30 MHz?
A. A certificate from the FCC showing that you have notified them that you will be using the HF bands
B. A certificate showing that you have attended a class in HF communications
C. A Certificate of Successful Completion of Examination showing that you have passed a Morse code exam
D. No special proof is required

T1D06 (C) [97.509a]
What is a Volunteer Examiner (VE)?
A. A certified instructor who volunteers to examine amateur teaching manuals
B. An FCC employee who accredits volunteers to administer amateur license exams
C. An amateur, accredited by one or more VECs, who volunteers to administer amateur license exams
D. An amateur, registered with the Electronic Industries Association, who volunteers to examine amateur station equipment

T1D07 (C) [97.503b1]
What minimum examinations must you pass for a Technician amateur license?
A. A written exam, Element 1 and a 5 WPM code exam, Element 2
B. A 5 WPM code exam, Element 1 and a written exam, Element 3
C. A single 35 question multiple choice written exam, Element 2
D. A written exam, Element 2 and a 5 WPM code exam, Element 4

T1D08 (D) [VE Instructions]
How may an Element 1 exam be administered to an applicant with a physical disability?
A. It may be skipped if a doctor signs a statement saying the applicant is too disabled to pass the exam
B. By holding an open book exam
C. By lowering the exam's pass rate to 50 percent correct
D. By using a vibrating surface or flashing light

T1D09 (A) [97.503a]
What is the purpose of the Element 1 examination?
A. To test Morse code comprehension at 5 words-per-minute
B. To test knowledge of block diagrams
C. To test antenna-building skills
D. To test knowledge of rules and regulations

T1D10 (A) [97.505(A) (6)]
If a Technician class licensee passes only the 5 words-per-minute Morse code test at an exam session, how long will this credit be valid for license upgrade purposes?
A. 365 days
B. Until the current license expires
C. Indefinitely
D. Until two years following the expiration of the current license

T1D11 (D) [97.505(A) (3), (5) and (7)]
Which of the following would confer credit for examination Element 1?
A. A Novice class license that expired more than two years ago
B. A current or expired Technician Plus license
C. A commercial radiotelegraph license or permit that is current or expired for less than five years
D. All of these are correct

T1E Amateur station call sign systems including Sequential, Vanity and Special Event; ITU Regions; Call sign formats.

T1E01 (C)
Which of the following call signs is a valid US amateur call?
A. UZ4FWD
B. KBL7766
C. KB3TMJ
D. VE3BKJ

T1E02 (B)
What letters must be used for the first letter in

US amateur call signs?
A. K, N, U and W
B. A, K, N and W
C. A, B, C and D
D. A, N, V and W

T1E03 (D)
What numbers are normally used in US amateur call signs?
A. Any two-digit number, 10 through 99
B. Any two-digit number, 22 through 45
C. A single digit, 1 though 9
D. A single digit, 0 through 9

T1E04 (B)
In which ITU region is Alaska?
A. ITU Region 1
B. ITU Region 2
C. ITU Region 3
D. ITU Region 4

T1E05 (C)
In which ITU region is Guam?
A. ITU Region 1
B. ITU Region 2
C. ITU Region 3
D. ITU Region 4

T1E06 (B) [97.119a]
What must you transmit to identify your amateur station?
A. Your "handle"
B. Your call sign
C. Your first name and your location
D. Your full name

T1E07 (A) [97.19]
How might you obtain a call sign made up of your initials?
A. Under the vanity call sign program
B. In a sequential call sign program
C. In the special event call sign program
D. There is no provision for choosing a call sign

T1E08 (A) [97.21(A) (3)(ii)]
How may an Amateur Radio licensee change his call sign without applying for a vanity call?
A. By requesting a systematic call sign change on an NCVEC Form 605
B. Paying a Volunteer Examiner team to process a call sign change request
C. By requesting a specific new call sign on an NCVEC Form 605 and sending it to the FCC in Gettysburg, PA
D. Contacting the FCC ULS database using the Internet to request a call sign change

T1E09 (B) [97.17b2]
How may an Amateur Radio club obtain a station call sign?
A. You must apply directly to the FCC in Gettysburg, PA
B. You must apply through a Club Station Call Sign Administrator
C. You must submit FCC Form 605 to FCC in Washington, DC
D. You must notify VE team on NCVEC Form 605

T1E10 (C)
Amateurs of which license classes are eligible to apply for temporary use of a
1-by-1 format Special Event call sign?
A. Only Amateur Extra class amateurs
B. 1-by-1 format call signs are not authorized in the US Amateur Service
C. Any FCC-licensed amateur
D. Only trustees of Amateur Radio clubs

T1E11 (C) [97.17d]
How does the FCC issue new Amateur Radio call signs?
A. By call sign district in random order
B. The applicant chooses a call sign no one else is using
C. By ITU prefix letter(s), call sign district numeral and a suffix in strict alphabetic order
D. The Volunteer Examiners who gave the exams choose a call sign no one else is using

T1E12 (D)
Which station call sign format groups are available to Technician Class amateur
radio operators?
A. Group A
B. Group B

C. Only Group C
D. Group C and D

SUBELEMENT T2 — Methods of Communication [2 Exam Questions — 2 Groups] T2A How Radio Works; Electromagnetic spectrum; Magnetic/Electric Fields; Nature of Radio Waves; Wavelength; Frequency; Velocity; AC Sine wave/Hertz; Audio and Radio frequency.

T2A01 (A)
What happens to a signal's wavelength as its frequency increases?
A. It gets shorter
B. It gets longer
C. It stays the same
D. It disappears

T2A02 (C)
How does the frequency of a harmonic compare to the desired transmitting
frequency?
A. It is slightly more than the desired frequency
B. It is slightly less than the desired frequency
C. It is exactly two, or three, or more times the desired frequency
D. It is much less than the desired frequency

T2A03 (B)
What does 60 hertz (Hz) mean?
A. 6000 cycles per second
B. 60 cycles per second
C. 6000 meters per second
D. 60 meters per second

T2A04 (C)
What is the name for the distance an AC signal travels during one complete
cycle?
A. Wave speed
B. Waveform
C. Wavelength
D. Wave spread

T2A05 (A)
What is the fourth harmonic of a 50.25 MHz signal?
A. 201.00 MHz
B. 150.75 MHz
C. 251.50 MHz
D. 12.56 MHz

T2A06 (C)
What is a radio frequency wave?
A. Wave disturbances that take place at less than 10 times per second
B. Electromagnetic oscillations or cycles that repeat between 20 and 20,000 times per second
C. Electromagnetic oscillations or cycles that repeat more than 20,000 times per second
D. None of these answers are correct

T2A07 (B)
What is an audio-frequency signal?
A. Wave disturbances that cannot be heard by the human ear
B. Electromagnetic oscillations or cycles that repeat between 20 and 20,000 times per second
C. Electromagnetic oscillations or cycles that repeat more than 20,000 times per second
D. Electric energy that is generated at the front end of an AM or FM radio receiver

T2A08 (D)
In what radio-frequency range do amateur 2-meter communications take place?
A. UHF, Ultra High Frequency range
B. MF, Medium Frequency range
C. HF, High Frequency range
D. VHF, Very High Frequency range

T2A09 (A)
Which of the following choices is often used to identify a particular radio wave?
A. The frequency or the wavelength of the wave
B. The length of the magnetic curve of wave
C. The time it takes for the wave to travel a certain distance
D. The free-spare impedance of the wave

T2A10 (D)
How is a radio frequency wave identified?
A. By its wavelength, the length of a single radio

cycle from peak to peak
B. By its corresponding frequency
C. By the appropriate radio band in which it is transmitted or received
D. All of these choices are correct

T2A11 (A)
How fast does a radio wave travel through space (in a vacuum)?
A. At the speed of light
B. At the speed of sound
C. Its speed is inversely proportional to its wavelength
D. Its speed increases as the frequency increases

T2A12 (B)
What is the standard unit of frequency measurement?
A. A megacycle
B. A hertz
C. One thousand cycles per second
D. EMF, electromagnetic force

T2A13 (A)
What is the basic principle of radio communications?
A. A radio wave is combined with an information signal and is transmitted; a receiver separates the two
B. A transmitter separates information to be received from a radio wave
C. A DC generator combines some type of information into a carrier wave so that it may travel through space
D. The peak-to-peak voltage of a transmitter is varied by the sidetone and modulated by the receiver

T2A14 (B)
How is the wavelength of a radio wave related to its frequency?
A. Wavelength gets longer as frequency increases
B. Wavelength gets shorter as frequency increases
C. There is no relationship between wavelength and frequency
D. The frequency depends on the velocity of the radio wave, but the wavelength depends on the bandwidth of the signal

T2A15 (D)
What term means the number of times per second that an alternating current flows back and forth?
A. Pulse rate
B. Speed
C. Wavelength
D. Frequency

T2A16 (A)
What is the basic unit of frequency?
A. The hertz
B. The watt
C. The ampere
D. The ohm

T2B Frequency privileges granted to Technician class operators; Amateur service bands; Emission types and designators; Modulation principles; AM/FM/Single sideband/upper-lower, international Morse code (CW), RTTY, packet radio and data emission types; Full quieting.

T2B01 (B) [97.301e]
What are the frequency limits of the 80-meter band in ITU Region 2 for
Technician class licensees who have passed a Morse code exam?
A. 3500 - 4000 kHz
B. 3675 - 3725 kHz
C. 7100 - 7150 kHz
D. 7000 - 7300 kHz

T2B02 (C) [97.301e]
What are the frequency limits of the 10-meter band in ITU Region 2 for
Technician class licensees who have passed a Morse code exam?
A. 28.000 - 28.500 MHz
B. 28.100 - 29.500 MHz
C. 28.100 - 28.500 MHz
D. 29.100 - 29.500 MHz

T2B03 (C) [97.3c2]
What name does the FCC use for telemetry, telecommand or computer communications emissions?

A. CW
B. Image
C. Data
D. RTTY

T2B04 (C)
What does "connected" mean in a packet-radio link?
A. A telephone link is working between two stations
B. A message has reached an amateur station for local delivery
C. A transmitting station is sending data to only one receiving station; it replies that the data is being received correctly
D. A transmitting and receiving station are using a digipeater, so no other contacts can take place until they are finished

T2B05 (D) [97.305, 97.307f9]
What emission types are Technician control operators who have passed a Morse code exam allowed to use from 7100 to 7150 kHz in ITU Region 2?
A. CW and data
B. Phone
C. Data only
D. CW only

T2B06 (C) [97.305, 97.307f10]
What emission types are Technician control operators who have passed a Morse code exam allowed to use on frequencies from 28.3 to 28.5 MHz?
A. All authorized amateur emission privileges
B. CW and data
C. CW and single-sideband phone
D. Data and phone

T2B07 (D) [97.305]
What emission types are Technician control operators allowed to use on the amateur 1.25-meter band in ITU Region 2?
A. Only CW and phone
B. Only CW and data
C. Only data and phone
D. All amateur emission privileges authorized for use on the band

T2B08 (B)
What term describes the process of combining an information signal with a radio signal?
A. Superposition
B. Modulation
C. Demodulation
D. Phase-inversion

T2B09 (D)
What is the name of the voice emission most used on VHF/UHF repeaters?
A. Single-sideband phone
B. Pulse-modulated phone
C. Slow-scan phone
D. Frequency-modulated phone

T2B10 (C)
What does the term "phone transmissions" usually mean?
A. The use of telephones to set up an amateur contact
B. A phone patch between Amateur Radio and the telephone system
C. AM, FM or SSB voice transmissions by radiotelephony
D. Placing the telephone handset near a transceiver's microphone and speaker to relay a telephone call

T2B11 (A)
Which sideband is commonly used for 10-meter phone operation?
A. Upper sideband
B. Lower sideband
C. Amplitude-compandored sideband
D. Double sideband

T2B12 (C) [97.313c2]
What is the most transmitter power a Technician control operator with telegraphy credit may use on the 10-meter band?
A. 5 watts PEP output
B. 25 watts PEP output
C. 200 watts PEP output
D. 1500 watts PEP output

T2B13 (D) [97.3c5]
What name does the FCC use for voice emissions?

A. RTTY
B. Data
C. CW
D. Phone

T2B14 (B) [97.305c]
What emission privilege is permitted a Technician class operator in the 219 MHz - 220 MHz frequency range?
A. Slow-scan television
B. Point-to-point digital message forwarding
C. FM voice
D. Fast-scan television

T2B15 (A)
Which sideband is normally used for VHF/UHF SSB communications?
A. Upper sideband
B. Lower sideband
C. Double sideband
D. Double sideband, suppressed carrier

T2B16 (A)
Which of the following descriptions is used to describe a good signal through a repeater?
A. Full quieting
B. Over deviation
C. Breaking up
D. Readability zero

T2B17 (B)
What is the typical bandwidth of PSK31 digital communications?
A. 500 kHz
B. 31 Hz
C. 5 MHz
D. 600 kHz

T2B18 (D)
What emissions do a transmitter using a reactance modulator produce?
A. CW
B. Test
C. Single-sideband, suppressed-carrier phone
D. Phase-modulated phone

T2B19 (C)
What other emission does phase modulation most resemble?
A. Amplitude modulation
B. Pulse modulation
C. Frequency modulation
D. Single-sideband modulation

SUBELEMENT T3 - Radio Phenomena [2 Exam Questions - 2 Groups]
T3A How a radio signal travels; Atmosphere/troposphere/ionosphere and ionized layers; Skip distance; Ground (surface)/sky (space) waves; Single/multihop; Path; Ionospheric absorption; Refraction.

T3A01 (D)
What is the name of the area of the atmosphere that makes long-distance radio communications possible by bending radio waves?
A. Troposphere
B. Stratosphere
C. Magnetosphere
D. Ionosphere

T3A02 (B)
Which ionospheric region is closest to the Earth?
A. The A region
B. The D region
C. The E region
D. The F region

T3A03 (D)
Which region of the ionosphere is mainly responsible for absorbing MF/HF radio signals during the daytime?
A. The F2 region
B. The F1 region
C. The E region
D. The D region

T3A04 (D)
Which region of the ionosphere is mainly responsible for long-distance sky-wave radio communications?
A. D region
B. E region
C. F1 region
D. F2 region

T3A05 (D)
When a signal travels along the surface of the Earth, what is this called?
A. Sky-wave propagation
B. Knife-edge diffraction
C. E-region propagation
D. Ground-wave propagation

T3A06 (C)
What type of solar radiation is most responsible for ionization in the outer atmosphere?
A. Thermal
B. Non-ionized particle
C. Ultraviolet
D. Microwave

T3A07 (C)
What is the usual cause of sky-wave propagation?
A. Signals are reflected by a mountain
B. Signals are reflected by the Moon
C. Signals are bent back to Earth by the ionosphere
D. Signals are retransmitted by a repeater

T3A08 (B)
What type of propagation has radio signals bounce several times between Earth and the ionosphere as they travel around the Earth?
A. Multiple bounce
B. Multi-hop
C. Skip
D. Pedersen propagation

T3A09 (A)
What effect does the D region of the ionosphere have on lower-frequency HF signals in the daytime?
A. It absorbs the signals
B. It bends the radio waves out into space
C. It refracts the radio waves back to earth
D. It has little or no effect on 80-meter radio waves

T3A10 (C)
How does the signal loss for a given path through the troposphere vary with frequency?
A. There is no relationship
B. The path loss decreases as the frequency increases
C. The path loss increases as the frequency increases
D. There is no path loss at all

T3A11 (A)
When a signal is returned to Earth by the ionosphere, what is this called?
A. Sky-wave propagation
B. Earth-Moon-Earth propagation
C. Ground-wave propagation
D. Tropospheric propagation

T3A12 (B)
How does the range of sky-wave propagation compare to ground-wave propagation?
A. It is much shorter
B. It is much longer
C. It is about the same
D. It depends on the weather

T3B HF vs. VHF vs. UHF characteristics; Types of VHF-UHF propagation; Daylight and seasonal variations; Tropospheric ducting; Line of sight; Maximum usable frequency (MUF); Sunspots and sunspot Cycle, Characteristics of different bands.

T3B01 (A)
When a signal travels in a straight line from one antenna to another, what is this called?
A. Line-of-sight propagation
B. Straight line propagation
C. Knife-edge diffraction
D. Tunnel ducting

T3B02 (C)
What can happen to VHF or UHF signals going towards a metal-framed building?
A. They will go around the building
B. They can be bent by the ionosphere
C. They can be reflected by the building
D. They can be polarized by the building's mass

T3B03 (C)
Ducting occurs in which region of the atmosphere?
A. F2

B. Ecosphere
C. Troposphere
D. Stratosphere

T3B04 (B)
What causes VHF radio waves to be propagated several hundred miles over oceans?
A. A polar air mass
B. A widespread temperature inversion
C. An overcast of cirriform clouds
D. A high-pressure zone

T3B05 (B)
In which of the following frequency ranges does sky-wave propagation least often occur?
A. LF
B. UHF
C. HF
D. VHF

T3B06 (A)
Why should local amateur communications use VHF and UHF frequencies instead of HF frequencies?
A. To minimize interference on HF bands capable of long-distance communication
B. Because greater output power is permitted on VHF and UHF
C. Because HF transmissions are not propagated locally
D. Because signals are louder on VHF and UHF frequencies

T3B07 (A)
How does the number of sunspots relate to the amount of ionization in the ionosphere?
A. The more sunspots there are, the greater the ionization
B. The more sunspots there are, the less the ionization
C. Unless there are sunspots, the ionization is zero
D. Sunspots do not affect the ionosphere

T3B08 (C)
How long is an average sunspot cycle?
A. 2 years
B. 5 years
C. 11 years
D. 17 years

T3B09 (B)
Which of the following frequency bands is most likely to experience summertime sporadic-E propagation?
A. 23 centimeters
B. 6 meters
C. 70 centimeters
D. 1.25 meters

T3B10 (A)
Which of the following emission modes are considered to be weak-signal modes and have the greatest potential for DX contacts?
A. Single sideband and CW
B. Packet radio and RTTY
C. Frequency modulation
D. Amateur television

T3B11 (D)
What is the condition of the ionosphere above a particular area of the Earth just before local sunrise?
A. Atmospheric attenuation is at a maximum
B. The D region is above the E region
C. The E region is above the F region
D. Ionization is at a minimum

T3B12 (A)
What happens to signals that take off vertically from the antenna and are higher in frequency than the critical frequency?
A. They pass through the ionosphere
B. They are absorbed by the ionosphere
C. Their frequency is changed by the ionosphere to be below the maximum usable frequency
D. They are reflected back to their source

T3B13 (A)
In relation to sky-wave propagation, what does the term "maximum usable frequency" (MUF) mean?
A. The highest frequency signal that will reach its intended destination
B. The lowest frequency signal that will reach its intended destination

C. The highest frequency signal that is most absorbed by the ionosphere
D. The lowest frequency signal that is most absorbed by the ionosphere

SUBELEMENT T4 — Station Licensee Duties [3 Exam Questions — 3 Groups]
T4A Correct name and mailing address on station license grant; Places from where station is authorized to transmit; Selecting station location; Antenna structure location; Stations installed aboard ship or aircraft.

T4A01 (C) [97.11a]
When may you operate your amateur station aboard a cruise ship?
A. At any time
B. Only while the ship is not under power
C. Only with the approval of the master of the ship and not using the ship's radio quipment
D. Only when you have written permission from the cruise line and only using the ship's radio equipment

T4A02 (D)
When may you operate your amateur station somewhere in the US besides the address listed on your license?
A. Only during times of emergency
B. Only after giving proper notice to the FCC
C. During an emergency or an FCC-approved emergency practice
D. Whenever you want to

T4A03 (C) [97.23]
What penalty may the FCC impose if you fail to provide your correct mailing address?
A. There is no penalty if you do not provide the correct address
B. You are subject to an administrative fine
C. Your amateur license could be revoked
D. You may only operate from your address of record

T4A04 (A)
Under what conditions may you transmit from a location different from the address printed on your amateur license?
A. If the location is under the control of the FCC, whenever the FCC Rules allow
B. If the location is outside the United States, only for a time period of less than 90 days
C. Only when you have written permission from the FCC Engineer in Charge
D. Never; you may only operate at the location printed on your license

T4A05 (B) [97.23]
Why must an amateur operator have a current US postal mailing address?
A. So the FCC has a record of the location of each amateur station
B. To follow the FCC rules and so the licensee can receive mail from the FCC
C. Because all US amateurs must be US residents
D. So the FCC can publish a call-sign directory

T4A06 (B) [97.21a1]
What is one way to notify the FCC if your mailing address changes?
A. Fill out an FCC Form 605 using your new address, attach a copy of your license, and mail it to your local FCC Field Office
B. Fill out an FCC Form 605 using your new address, attach a copy of your license, and mail it to the FCC office in Gettysburg, PA
C. Call your local FCC Field Office and give them your new address over the phone
D. Call the FCC office in Gettysburg, PA, and give them your new address over the phone

T4A07 (A) [97.15(A)]
What do FCC rules require you to do if you plan to erect an antenna whose height exceeds 200 feet?
A. Notify the Federal Aviation Administration and register with the FCC
B. FCC rules prohibit antenna structures above 200 feet
C. Alternating sections of the supporting structure must be painted international airline orange and white
D. The antenna structure must be approved by the FCC and DOD

T4A08 (D) [97.13c] [OET Bulletin 65 Supplement B] ["RF Exposure and You", W1RFI]
Which of the following is NOT an important consideration when selecting a location for a transmitting antenna?
A. Nearby structures
B. Height above average terrain
C. Distance from the transmitter location
D. Polarization of the feed line

T4A09 (B) [97.15b]
What is the height restriction the FCC places on Amateur Radio Service antenna structures without registration with the FCC and FAA?
A. There is no restriction by the FCC
B. 200 feet
C. 300 feet
D. As permitted by PRB-1

T4A10 (C) [97.11a]
When may you operate your amateur station aboard an aircraft?
A. At any time
B. Only while the aircraft is on the ground
C. Only with the approval of the pilot in command and not using the aircraft's radio equipment
D. Only when you have written permission from the airline and only using the aircraft's radio equipment

T4B Designation of control operator; FCC presumption of control operator; Physical control of station apparatus; Control point; Immediate station control; Protecting against unauthorized transmissions; Station records; FCC Inspection; Restricted operation.

T4B01 (C) [97.3a12]
What is the definition of a control operator of an amateur station?
A. Anyone who operates the controls of the station
B. Anyone who is responsible for the station's equipment
C. Any licensed amateur operator who is responsible for the station's transmissions
D. The amateur operator with the highest class of license who is near the controls of the station

T4B02 (D) [97.3a12]
What is the FCC's name for the person esponsible for the transmissions from an amateur station?
A. Auxiliary operator
B. Operations coordinator
C. Third-party operator
D. Control operator

T4B03 (C) [97.7]
When must an amateur station have a control operator?
A. Only when training another amateur
B. Whenever the station receiver is operated
C. Whenever the station is transmitting
D. A control operator is not needed

T4B04 (B) [97.3a13]
What is the term for the location at which the control operator function is performed?
A. The operating desk
B. The control point
C. The station location
D. The manual control location

T4B05 (D) [97.3a13]
What is the control point of an amateur station?
A. The on/off switch of the transmitter
B. The input/output port of a packet controller
C. The variable frequency oscillator of a transmitter
D. The location at which the control operator function is performed

T4B06 (D) [97.3a12]
When you operate your transmitting equipment alone, what is your official designation?
A. Engineer in Charge
B. Commercial radio operator
C. Third party
D. Control operator

T4B07 (A) [97.103b]
When does the FCC assume that you authorize

transmissions with your call sign as the control operator?
A. At all times
B. Only in the evening hours
C. Only when operating third party traffic
D. Only when operating as a reciprocal operating station

T4B08 (C) [97.3a13]
What is the name for the operating position where the control operator has full control over the transmitter?
A. Field point
B. Auxiliary point
C. Control point
D. Access point

T4B09 (B) [97.103c]
When is the FCC allowed to conduct an inspection of your amateur station?
A. Only on weekends
B. At any time
C. Never, the FCC does not inspect stations
D. Only during daylight hours

T4B10 (C) [97.5d]
How many transmitters may an amateur licensee control at the same time?
A. Only one
B. No more than two
C. Any number
D. Any number, as long as they are transmitting in different bands

T4B11 (A) [97.121]
If you have been informed that your Amateur Radio station causes interference to nearby radio or television broadcast receivers of good engineering design, what operating restrictions can FCC rules impose on your station?
A. Require that you discontinue operation on frequencies causing interference during certain evening hours and on Sunday morning (local time)
B. Relocate your station or reduce your transmitter's output power
C. Nothing, unless the FCC conducts an investigation of the interference problem and issues a citation
D. Reduce antenna height so as to reduce the area affected by the interference

T4B12 (B)
How could you best keep unauthorized persons from using your amateur station at home?
A. Use a carrier-operated relay in the main power line
B. Use a key-operated on/off switch in the main power line
C. Put a "Danger - High Voltage" sign in the station
D. Put fuses in the main power line

T4B13 (A)
How could you best keep unauthorized persons from using a mobile amateur station in your car?
A. Disconnect the microphone when you are not using it
B. Put a "do not touch" sign on the radio
C. Turn the radio off when you are not using it
D. Tune the radio to an unused frequency when you are done using it

T4C Providing public service; emergency and disaster communications; Distress calling; Emergency drills and communications; Purpose of RACES.

T4C01 (D) [97.405a]
If you hear a voice distress signal on a frequency outside of your license privileges, what are you allowed to do to help the station in distress?
A. You are NOT allowed to help because the frequency of the signal is outside your privileges
B. You are allowed to help only if you keep your signals within the nearest frequency band of your privileges
C. You are allowed to help on a frequency outside your privileges only if you use international Morse code
D. You are allowed to help on a frequency outside your privileges in any way possible

T4C02 (C) [97.403]
When may you use your amateur station to

transmit an "SOS" or "MAYDAY"?
A. Never
B. Only at specific times (at 15 and 30 minutes after the hour)
C. In a life- or property-threatening emergency
D. When the National Weather Service has announced a severe weather watch

T4C03 (A) [97.401a]
If a disaster disrupts normal communication systems in an area where the FCC regulates the amateur service, what kinds of transmissions may stations make?
A. Those that are necessary to meet essential communication needs and facilitate relief actions
B. Those that allow a commercial business to continue to operate in the affected area
C. Those for which material compensation has been paid to the amateur operator for delivery into the affected area
D. Those that are to be used for program production or newsgathering for broadcasting purposes

T4C04 (C) [97.401c]
What information is included in an FCC declaration of a temporary state of communication emergency?
A. A list of organizations authorized to use radio communications in the affected area
B. A list of amateur frequency bands to be used in the affected area
C. Any special conditions and special rules to be observed during the emergency
D. An operating schedule for authorized amateur emergency stations

T4C05 (D)
If you are in contact with another station and you hear an emergency call for help on your frequency, what should you do?
A. Tell the calling station that the frequency is in use
B. Direct the calling station to the nearest emergency net frequency
C. Call your local Civil Preparedness Office and inform them of the emergency
D. Stop your QSO immediately and take the emergency call

T4C06 (A)
What is the proper way to interrupt a repeater conversation to signal a distress call?
A. Say "BREAK" once, then your call sign
B. Say "HELP" as many times as it takes to get someone to answer
C. Say "SOS," then your call sign
D. Say "EMERGENCY" three times

T4C07 (B)
What is one reason for using tactical call signs such as "command post" or "weather center" during an emergency?
A. They keep the general public informed about what is going on
B. They are more efficient and help coordinate public-service communications
C. They are required by the FCC
D. They increase goodwill between amateurs

T4C08 (D)
What type of messages concerning a person's well being are sent into or out of a disaster area?
A. Routine traffic
B. Tactical traffic
C. Formal message traffic
D. Health and welfare traffic

T4C09 (B)
What are messages called that are sent into or out of a disaster area concerning the immediate safety of human life?
A. Tactical traffic
B. Emergency traffic
C. Formal message traffic
D. Health and welfare traffic

T4C10 (B)
Why is it a good idea to have a way to operate your amateur station without using commercial AC power lines?
A. So you may use your station while mobile
B. So you may provide communications in an emergency
C. So you may operate in contests where AC

power is not allowed
D. So you will comply with the FCC rules

T4C11 (C)
What is the most important accessory to have for a hand-held radio in an emergency?
A. An extra antenna
B. A portable amplifier
C. Several sets of charged batteries
D. A microphone headset for hands-free operation

T4C12 (C)
Which type of antenna would be a good choice as part of a portable HF amateur station that could be set up in case of an emergency?
A. A three-element quad
B. A three-element Yagi
C. A dipole
D. A parabolic dish

T4C13 (D)
How must you identify messages sent during a RACES drill?
A. As emergency messages
B. As amateur traffic
C. As official government messages
D. As drill or test messages

T4C14 (C)
With what organization must you register before you can participate in RACES drills?
A. A local Amateur Radio club
B. A local racing organization
C. The responsible civil defense organization
D. The Federal Communications Commission

SUBELEMENT T5 - Control Operator Duties [3 Exam Questions — 3 Groups]
T5A Determining operating privileges, Where control operator must be situated while station is locally or remotely controlled; Operating other amateur stations.

T5A01 (B) [97.105b]
If you are the control operator at the station of another amateur who has a higher-class license than yours, what operating privileges are you allowed?
A. Any privileges allowed by the higher license
B. Only the privileges allowed by your license
C. All the emission privileges of the higher license, but only the frequency privileges of your license
D. All the frequency privileges of the higher license, but only the emission privileges of your license

T5A02 (A)
Assuming you operate within your amateur license privileges, what restrictions apply to operating amateur equipment?
A. You may operate any amateur equipment
B. You may only operate equipment located at the address printed on your amateur license
C. You may only operate someone else's equipment if you first notify the FCC
D. You may only operate store-purchased equipment until you earn your Amateur Extra class license

T5A03 (A) [97.109b]
When an amateur station is transmitting, where must its control operator be, assuming the station is not under automatic control?
A. At the station's control point
B. Anywhere in the same building as the transmitter
C. At the station's entrance, to control entry to the room
D. Anywhere within 50 km of the station location

T5A04 (B)
Where will you find a detailed list of your operating privileges?
A. In the OET Bulletin 65 Index
B. In FCC Part 97
C. In your equipment's operating instructions
D. In Part 15 of the Code of Federal Regulations

T5A05 (A) [97.103a]
If you transmit from another amateur's station, who is responsible for its proper operation?
A. Both of you

B. The other amateur (the station licensee)
C. You, the control operator
D. The station licensee, unless the station records show that you were the control operator at the time

T5A06 (A) [97.105b]
If you let another amateur with a higher class license than yours control your station, what operating privileges are allowed?
A. Any privileges allowed by the higher license, as long as proper identification procedures are followed
B. Only the privileges allowed by your license
C. All the emission privileges of the higher license, but only the frequency privileges of your license
D. All the frequency privileges of the higher license, but only the emission privileges of your license

T5A07 (B) [97.105(B)]
If a Technician class licensee uses the station of a General class licensee, how may the Technician licensee operate?
A. Within the frequency limits of a General class license
B. Within the limits of a Technician class license
C. Only as a third party with the General class licensee as the control operator
D. A Technician class licensee may not operate a General class station

T5A08 (C) [97.109(D)]
What type of amateur station does not require the control operator to be present at the control point?
A. A locally controlled station
B. A remotely controlled station
C. An automatically controlled station
D. An earth station controlling a space station

T5A09 (B) [97.109b]
Why can't unlicensed persons in your family transmit using your amateur station if they are alone with your equipment?
A. They must not use your equipment without your permission
B. They must be licensed before they are allowed to be control operators
C. They must first know how to use the right abbreviations and Q signals
D. They must first know the right frequencies and emissions for transmitting

T5A10 (C)
If you own a dual-band mobile transceiver, what requirement must be met if you set it up to operate as a crossband repeater?
A. There is no special requirement if you are licensed for both bands
B. You must hold an Amateur Extra class license
C. There must be a control operator at the system's control point
D. Operating a crossband mobile system is not allowed

T5B Transmitter power standards; Interference to stations providing emergency communications; Station identification requirements.

T5B01 (C) [97.119a]
How often must an amateur station be identified?
A. At the beginning of a contact and at least every ten minutes after that
B. At least once during each transmission
C. At least every ten minutes during and at the end of a contact
D. At the beginning and end of each transmission

T5B02 (C) [97.119a]
What identification, if any, is required when two amateur stations end communications?
A. No identification is required
B. One of the stations must transmit both stations' call signs
C. Each station must transmit its own call sign
D. Both stations must transmit both call signs

T5B03 (B) [97.119a]
What is the longest period of time an amateur station can operate without transmitting its call sign?
A. 5 minutes
B. 10 minutes

C. 15 minutes
D. 30 minutes

T5B04 (A) [97.305a]
What emission type may always be used for station identification, regardless of the transmitting frequency?
A. CW
B. RTTY
C. MCW
D. Phone

T5B05 (D) [97.3b6]
What is the term for the average power supplied to an antenna transmission line during one RF cycle at the crest of the modulation envelope?
A. Peak transmitter power
B. Peak output power
C. Average radio-frequency power
D. Peak envelope power

T5B06 (A) [97.313c]
On which band(s) may a Technician licensee who has passed a Morse code exam use up to 200 watts PEP output power?
A. 80, 40, 15, and 10 meters
B. 80, 40, 20, and 10 meters
C. 1.25 meters
D. 23 centimeters

T5B07 (D) [97.313a]
What amount of transmitter power must amateur stations use at all times?
A. 25 watts PEP output
B. 250 watts PEP output
C. 1500 watts PEP output
D. The minimum legal power necessary to communicate

T5B08 (C) [97.119b2]
If you are using a language besides English to make a contact, what language must you use when identifying your station?
A. The language being used for the contact
B. The language being used for the contact, provided the US has a third-party communications agreement with that country
C. English
D. Any language of a country that is a member of the International Telecommunication Union

T5B09 (C)
If you are helping in a communications emergency that is being handled by a net control operator, how might you best minimize interference to the net once you have checked in?
A. Whenever the net frequency is quiet, announce your call sign and location
B. Move 5 kHz away from the net's frequency and use high power to ask for other emergency communications
C. Do not transmit on the net frequency until asked to do so by the net operator
D. Wait until the net frequency is quiet, then ask for any emergency traffic for your area

T5B10 (D) [97.215a]
What are the station identification requirements for an amateur transmitter used for telecommand (control) of model craft?
A. Once every ten minutes
B. Once every ten minutes, and at the beginning and end of each transmission
C. At the beginning and end of each transmission
D. Station identification is not required if the transmitter is labeled with the station licensee's name, address and call sign

T5B11 (B) [97.3a23]
Why is transmitting on a police frequency as a "joke" called harmful interference that deserves a large penalty?
A. It annoys everyone who listens
B. It blocks police calls that might be an emergency and interrupts police communications
C. It is in bad taste to communicate with non-amateurs, even as a joke
D. It is poor amateur practice to transmit outside the amateur bands

T5B12 (D) [97.303]
If you are using a frequency within a band

assigned to the amateur service on a secondary basis, and a station assigned to the primary service on that band causes interference, what action should you take?

A. Notify the FCC's regional Engineer in Charge of the interference
B. Increase your transmitter's power to overcome the interference
C. Attempt to contact the station and request that it stop the interference
D. Change frequencies; you may be causing harmful interference to the other station, in violation of FCC rules

T5C Authorized transmissions, Prohibited practices; Third party communications; Retransmitting radio signals; One way communications.

T5C01 (D) [97.119a]
If you answer someone on the air and then complete your communication without giving your call sign, what type of communication have you just conducted?

A. Test transmission
B. Tactical signal
C. Packet communication
D. Unidentified communication

T5C02 (A) [97.111(B) (3)]
What is one example of one-way communication that Technician class control operators are permitted by FCC rules?

A. Transmission for radio control of model craft
B. Use of amateur television for surveillance purposes
C. Retransmitting National Weather Service broadcasts
D. Use of Amateur Radio as a wireless microphone for a public address system

T5C03 (D) [97.11a2]
What kind of payment is allowed for third-party messages sent by an amateur station?

A. Any amount agreed upon in advance
B. Donation of repairs to amateur equipment
C. Donation of amateur equipment
D. No payment of any kind is allowed

T5C04 (A) [97.3a44]
What is the definition of third-party communications?

A. A message sent between two amateur stations for someone else
B. Public service communications for a political party
C. Any messages sent by amateur stations
D. A three-minute transmission to another amateur

T5C05 (D) [97.115a2]
When are third-party messages allowed to be sent to a foreign country?

A. When sent by agreement of both control operators
B. When the third party speaks to a relative
C. They are not allowed under any circumstances
D. When the US has a third-party agreement with the foreign country or the third party is qualified to be a control operator

T5C06 (A) [97.115b1]
If you let an unlicensed third party use your amateur station, what must you do at your station's control point?

A. You must continuously monitor and supervise the third-party's participation
B. You must monitor and supervise the communication only if contacts are made in countries that have no third-party communications agreement with the US
C. You must monitor and supervise the communication only if contacts are made on frequencies below 30 MHz
D. You must key the transmitter and make the station identification

T5C07 (B) [97.115c]
Besides normal identification, what else must a US station do when sending third-party communications internationally?

A. The US station must transmit its own call sign at the beginning of each communication, and at least every ten minutes after that
B. The US station must transmit both call signs at the end of each communication

C. The US station must transmit its own call sign at the beginning of each communication, and at least every five minutes after that
D. Each station must transmit its own call sign at the end of each transmission, and at least every five minutes after that

T5C08 (C) [97.113a4]
If an amateur pretends there is an emergency and transmits the word "MAYDAY," what is this called?
A. A traditional greeting in May
B. An emergency test transmission
C. False or deceptive signals
D. Nothing special; "MAYDAY" has no meaning in an emergency

T5C09 (C) [97.119a]
If an amateur transmits to test access to a repeater without giving any station identification, what type of communication is this called?
A. A test emission; no identification is required
B. An illegal unmodulated transmission
C. An illegal unidentified transmission
D. A non-communication; no voice is transmitted

T5C10 (C) [97.101d]
When may you deliberately interfere with another station's communications?
A. Only if the station is operating illegally
B. Only if the station begins transmitting on a frequency you are using
C. Never
D. You may expect, and cause, deliberate interference because it can't be helped during crowded band conditions

T5C11 (B) [97.3a22]
If an amateur repeatedly transmits on a frequency already occupied by a group of amateurs in a net operation, what type of interference is this called?
A. Break-in interference
B. Harmful or malicious interference
C. Incidental interference
D. Intermittent interference

T5C12 (B)
What device is commonly used to retransmit Amateur Radio signals?
A. A beacon
B. A repeater
C. A radio controller
D. A duplexer

SUBELEMENT T6 - Good Operating Practices [3 Exam Questions — 3 Groups]
T6A Calling another station; Calling CQ; Typical amateur service radio contacts; Courtesy and respect for others; Popular Q-signals; Signal reception reports; Phonetic alphabet for voice operations.

T6A01 (A) [97.119b2]
What is the advantage of using the International Telecommunication Union (ITU) phonetic alphabet when identifying your station?
A. The words are internationally recognized substitutes for letters
B. There is no advantage
C. The words have been chosen to represent Amateur Radio terms
D. It preserves traditions begun in the early days of Amateur Radio

T6A02 (A) [97.119b2]
What is one reason to avoid using "cute" phrases or word combinations to identify your station?
A. They are not easily understood by non-English-speaking amateurs
B. They might offend English-speaking amateurs
C. They do not meet FCC identification requirements
D. They might be interpreted as codes or ciphers intended to obscure the meaning of your identification

T6A03 (A)
What should you do before you transmit on any frequency?
A. Listen to make sure others are not using the frequency
B. Listen to make sure that someone will be able to hear you

C. Check your antenna for resonance at the selected frequency
D. Make sure the SWR on your antenna feed line is high enough

T6A04 (B)
How do you call another station on a repeater if you know the station's call sign?
A. Say "break, break 79," then say the station's call sign
B. Say the station's call sign, then identify your own station
C. Say "CQ" three times, then say the station's call sign
D. Wait for the station to call "CQ," then answer it

T6A05 (D)
What does RST mean in a signal report?
A. Recovery, signal strength, tempo
B. Recovery, signal speed, tone
C. Readability, signal speed, tempo
D. Readability, signal strength, tone

T6A06 (D)
What is the meaning of: "Your signal report is five nine plus 20 dB..."?
A. Your signal strength has increased by a factor of 100
B. Repeat your transmission on a frequency 20 kHz higher
C. The bandwidth of your signal is 20 decibels above linearity
D. A relative signal-strength meter reading is 20 decibels greater than strength 9

T6A07 (D)
What is the meaning of the procedural signal "CQ"?
A. Call on the quarter hour
B. New antenna is being tested (no station should answer)
C. Only the called station should transmit
D. Calling any station

T6A08 (C)
What is a QSL card in the amateur service?
A. A letter or postcard from an amateur pen pal
B. A Notice of Violation from the FCC
C. A written acknowledgment of communications between two amateurs
D. A postcard reminding you when your license will expire

T6A09 (C)
What is the correct way to call CQ when using voice?
A. Say "CQ" once, followed by "this is," followed by your call sign spoken three times
B. Say "CQ" at least five times, followed by "this is," followed by your call sign spoken once
C. Say "CQ" three times, followed by "this is," followed by your call sign spoken three times
D. Say "CQ" at least ten times, followed by "this is," followed by your call sign spoken once

T6A10 (D)
How should you answer a voice CQ call?
A. Say the other station's call sign at least ten times, followed by "this is," then your call sign at least twice
B. Say the other station's call sign at least five times phonetically, followed by "this is," then your call sign at least once
C. Say the other station's call sign at least three times, followed by "this is," then your call sign at least five times phonetically
D. Say the other station's call sign once, followed by "this is," then your call sign given phonetically

T6A11 (A)
What is the meaning of: "Your signal is full quieting..."?
A. Your signal is strong enough to overcome all receiver noise
B. Your signal has no spurious sounds
C. Your signal is not strong enough to be received
D. Your signal is being received, but no audio is being heard

T6A12 (B)
What is meant by the term "DX"?
A. Best regards
B. Distant station

C. Calling any station
D. Go ahead

T6A13 (B)
What is the meaning of the term "73"?
A. Long distance
B. Best regards
C. Love and kisses
D. Go ahead

T6B Occupied bandwidth for emission types; Mandated and voluntary band plans; CW operation.

T6B01 (C)
Which list of emission types is in order from the narrowest bandwidth to the widest bandwidth?
A. RTTY, CW, SSB voice, FM voice
B. CW, FM voice, RTTY, SSB voice
C. CW, RTTY, SSB voice, FM voice
D. CW, SSB voice, RTTY, FM voice

T6B02 (D)
What is the usual bandwidth of a single-sideband amateur signal?
A. 1 kHz
B. 2 kHz
C. Between 3 and 6 kHz
D. Between 2 and 3 kHz

T6B03 (C)
What is the usual bandwidth of a frequency-modulated amateur signal?
A. Less than 5 kHz
B. Between 5 and 10 kHz
C. Between 10 and 20 kHz
D. Greater than 20 kHz

T6B04 (B)
What is the usual bandwidth of a UHF amateur fast-scan television signal?
A. More than 6 MHz
B. About 6 MHz
C. About 3 MHz
D. About 1 MHz

T6B05 (A)
What name is given to an Amateur Radio station that is used to connect other amateur stations with the Internet?
A. A gateway
B. A repeater
C. A digipeater
D. FCC regulations prohibit such a station

T6B06 (A)
What is a band plan?
A. A voluntary guideline beyond the divisions established by the FCC for using different operating modes within an amateur band
B. A guideline from the FCC for making amateur frequency band allocations
C. A plan of operating schedules within an amateur band published by the FCC
D. A plan devised by a club to best use a frequency band during a contest

T6B07 (C)
At what speed should a Morse code CQ call be transmitted?
A. Only speeds below five WPM
B. The highest speed your keyer will operate
C. Any speed at which you can reliably receive
D. The highest speed at which you can control the keyer

T6B08 (A)
What is the meaning of the procedural signal "DE"?
A. "From" or "this is," as in "W0AIH DE KA9FOX"
B. "Directional Emissions" from your antenna
C. "Received all correctly"
D. "Calling any station"

T6B09 (B)
What is a good way to call CQ when using Morse code?
A. Send the letters "CQ" three times, followed by "DE," followed by your call sign sent once
B. Send the letters "CQ" three times, followed by "DE," followed by your call sign sent three times
C. Send the letters "CQ" ten times, followed by "DE," followed by your call sign sent twice

D. Send the letters "CQ" over and over until a station answers

T6B10 (B)
How should you answer a Morse code CQ call?
A. Send your call sign four times
B. Send the other station's call sign twice, followed by "DE," followed by your call sign twice
C. Send the other station's call sign once, followed by "DE," followed by your call sign four times
D. Send your call sign followed by your name, station location and a signal report

T6B11 (A)
What is the meaning of the procedural signal "K"?
A. "Any station transmit"
B. "All received correctly"
C. "End of message"
D. "Called station only transmit"

T6B12 (B)
What is one meaning of the Q signal "QRS"?
A. "Interference from static"
B. "Send more slowly"
C. "Send RST report"
D. "Radio station location is"

T6C TVI and RFI reduction and elimination, Band/Low/High pass filter, Out of band harmonic Signals, Spurious Emissions, Telephone Interference, Shielding, Receiver Overload.

T6C01 (C)
What is meant by receiver overload?
A. Too much voltage from the power supply
B. Too much current from the power supply
C. Interference caused by strong signals from a nearby source
D. Interference caused by turning the volume up too high

T6C02 (B)
What type of filter might be connected to an amateur HF transmitter to cut down on harmonic radiation?
A. A key-click filter
B. A low-pass filter
C. A high-pass filter
D. A CW filter

T6C03 (B)
What type of filter should be connected to a TV receiver as the first step in trying to prevent RF overload from an amateur HF station transmission?
A. Low-pass
B. High-pass
C. Band pass
D. Notch

T6C04 (C)
What effect might a break in a cable television transmission line have on amateur communications?
A. Cable lines are shielded and a break cannot affect amateur communications
B. Harmonic radiation from the TV receiver may cause the amateur transmitter to transmit off-frequency
C. TV interference may result when the amateur station is transmitting, or interference may occur to the amateur receiver
D. The broken cable may pick up very high voltages when the amateur station is transmitting

T6C05 (A)
If you are told that your amateur station is causing television interference, what should you do?
A. First make sure that your station is operating properly, and that it does not cause interference to your own television
B. Immediately turn off your transmitter and contact the nearest FCC office for assistance
C. Connect a high-pass filter to the transmitter output and a low-pass filter to the antenna-input terminals of the television
D. Continue operating normally, because you have no reason to worry about the interference

T6C06 (C)
If harmonic radiation from your transmitter is causing interference to television receivers in your neighborhood, who is responsible for taking care of the interference?
A. The owners of the television receivers are responsible
B. Both you and the owners of the television receivers share the responsibility
C. You alone are responsible, since your transmitter is causing the problem
D. The FCC must decide if you or the owners of the television receivers are responsible

T6C07 (D)
If signals from your transmitter are causing front-end overload in your neighbor's television receiver, who is responsible for taking care of the interference?
A. You alone are responsible, since your transmitter is causing the problem
B. Both you and the owner of the television receiver share the responsibility
C. The FCC must decide if you or the owner of the television receiver are responsible
D. The owner of the television receiver is responsible

T6C08 (A)
What circuit blocks RF energy above and below certain limits?
A. A band-pass filter
B. A high-pass filter
C. An input filter
D. A low-pass filter

T6C09 (D)
If someone tells you that signals from your hand-held transceiver are interfering with other signals on a frequency near yours, what may be the cause?
A. You may need a power amplifier for your hand-held
B. Your hand-held may have chirp from weak batteries
C. You may need to turn the volume up on your hand-held
D. Your hand-held may be transmitting spurious emissions

T6C10 (B)
What may happen if an SSB transmitter is operated with the microphone gain set too high?
A. It may cause digital interference to computer equipment
B. It may cause splatter interference to other stations operating near its frequency
C. It may cause atmospheric interference in the air around the antenna
D. It may cause interference to other stations operating on a higher frequency band

T6C11 (D)
What may cause a buzzing or hum in the signal of an HF transmitter?
A. Using an antenna that is the wrong length
B. Energy from another transmitter
C. Bad design of the transmitter's RF power output circuit
D. A bad filter capacitor in the transmitter's power supply

T6C12 (C) (Reference: FCC CIB Telephone Interference Bulletin)
What is the major cause of telephone interference?
A. The telephone ringer is inadequate
B. Tropospheric ducting at UHF frequencies
C. The telephone was not equipped with interference protection when it was manufactured.
D. Improper location of the telephone in the home

SUBELEMENT T7 Basic Communications Electronics [3 Exam Questions — 3 Groups]
T7A Fundamentals of electricity; AC/DC power; units and definitions of current, voltage, resistance, inductance, capacitance and impedance; Rectification; Ohm's Law principle (simple math); Decibel; Metric system and prefixes (e.g., pico, nano, micro, milli, deci, centi, kilo, mega, giga).

T7A01 (D)
What is the name for the flow of electrons in an electric circuit?

A. Voltage
B. Resistance
C. Capacitance
D. Current

T7A02 (B)
What is the name of a current that flows only in one direction?
A. An alternating current
B. A direct current
C. A normal current
D. A smooth current

T7A03 (A)
What is the name of a current that flows back and forth, first in one direction, then in the opposite direction?
A. An alternating current
B. A direct current
C. A rough current
D. A steady state current

T7A04 (B)
What is the basic unit of electrical power?
A. The ohm
B. The watt
C. The volt
D. The ampere

T7A05 (C)
What is the basic unit of electric current?
A. The volt
B. The watt
C. The ampere
D. The ohm

T7A06 (A)
How much voltage does an automobile battery usually supply?
A. About 12 volts
B. About 30 volts
C. About 120 volts
D. About 240 volts

T7A07 (D)
What limits the current that flows through a circuit for a particular applied DC voltage?
A. Reliance
B. Reactance
C. Saturation
D. Resistance

T7A08 (D)
What is the basic unit of resistance?
A. The volt
B. The watt
C. The ampere
D. The ohm

T7A09 (C)
What is the basic unit of inductance?
A. The coulomb
B. The farad
C. The henry
D. The ohm

T7A10 (A)
What is the basic unit of capacitance?
A. The farad
B. The ohm
C. The volt
D. The henry

T7A11 (B)
Which of the following circuits changes an alternating current signal into a varying direct current signal?
A. Transformer
B. Rectifier
C. Amplifier
D. Director

T7A12 (A)
What formula shows how voltage, current and resistance relate to each other in an electric circuit?
A. Ohm's Law
B. Kirchhoff's Law
C. Ampere's Law
D. Tesla's Law

T7A13 (C)
If a current of 2 amperes flows through a 50-ohm resistor, what is the voltage across the resistor?
A. 25 volts

B. 52 volts
C. 100 volts
D. 200 volts

T7A14 (B)
If a 100-ohm resistor is connected to 200 volts, what is the current through the resistor?
A. 1 ampere
B. 2 amperes
C. 300 amperes
D. 20,000 amperes

T7A15 (B)
If a current of 3 amperes flows through a resistor connected to 90 volts, what is the resistance?
A. 3 ohms
B. 30 ohms
C. 93 ohms
D. 270 ohms

T7A16 (B)
If you increase your transmitter output power from 5 watts to 10 watts, what decibel (dB) increase does that represent?
A. 2 dB
B. 3 dB
C. 5 dB
D. 10 dB

T7A17 (C)
If an ammeter marked in amperes is used to measure a 3000-milliampere current, what reading would it show?
A. 0.003 amperes
B. 0.3 amperes
C. 3 amperes
D. 3,000,000 amperes

T7A18 (C)
How many hertz are in a kilohertz?
A. 10
B. 100
C. 1000
D. 1,000,000

T7A19 (C)
If a dial marked in megahertz shows a reading of 3.525 MHz, what would it show if it were marked in kilohertz?
A. 0.003525 kHz
B. 35.25 kHz
C. 3525 kHz
D. 3,525,000 kHz

T7A20 (B)
How many microfarads is 1,000,000 picofarads?
A. 0.001 microfarads
B. 1 microfarad
C. 1000 microfarads
D. 1,000,000,000 microfarads

T7A21 (B)
If you have a hand-held transceiver with an output of 500 milliwatts, how many watts would this be?
A. 0.02
B. 0.5
C. 5
D. 50

T7B Basic electric circuits; Analog vs. digital communications; Audio/RF signal; Amplification.

T7B01 (A)
What type of electric circuit uses signals that can vary continuously over a certain range of voltage or current values?
A. An analog circuit
B. A digital circuit
C. A continuous circuit
D. A pulsed modulator circuit

T7B02 (B)
What type of electric circuit uses signals that have voltage or current values only in specific steps over a certain range?
A. An analog circuit
B. A digital circuit
C. A step modulator circuit
D. None of these choices is correct

T7B03 (C)
Which of the following is an example of an analog communications method?
A. Morse code (CW)

B. Packet Radio
C. Frequency-modulated (FM) voice
D. PSK31

T7B04 (D)
Which of the following is an example of a digital communications method?
A. Single-sideband (SSB) voice
B. Amateur Television (ATV)
C. FM voice
D. Radioteletype (RTTY)

T7B05 (B)
Most humans can hear sounds in what frequency range?
A. 0 - 20 Hz
B. 20 - 20,000 Hz
C. 200 - 200,000 Hz
D. 10,000 - 30,000 Hz

T7B06 (B)
Why do we call electrical signals in the frequency range of 20 Hz to 20,000 Hz audio frequencies?
A. Because the human ear cannot sense anything in this range
B. Because the human ear can sense sounds in this range
C. Because this range is too low for radio energy
D. Because the human ear can sense radio waves in this range

T7B07 (C)
What is the lowest frequency of electrical energy that is usually known as a radio frequency?
A. 20 Hz
B. 2,000 Hz
C. 20,000 Hz
D. 1,000,000 Hz

T7B08 (B)
Electrical energy at a frequency of 7125 kHz is in what frequency range?
A. Audio
B. Radio
C. Hyper
D. Super-high

T7B09 (C)
If a radio wave makes 3,725,000 cycles in one second, what does this mean?
A. The radio wave's voltage is 3725 kilovolts
B. The radio wave's wavelength is 3725 kilometers
C. The radio wave's frequency is 3725 kilohertz
D. The radio wave's speed is 3725 kilometers per second

T7B10 (A)
Which component can amplify a small signal using low voltages?
A. A PNP transistor
B. A variable resistor
C. An electrolytic capacitor
D. A multiple-cell battery

T7B11 (C)
Which component can amplify a small signal but normally uses high voltages?
A. A transistor
B. An electrolytic capacitor
C. A vacuum tube
D. A multiple-cell battery

T7C Concepts of Resistance/resistor; Capacitor/capacitance; Inductor/Inductance; Conductor/Insulator; Diode; Transistor; Semiconductor devices; Electrical functions of and schematic symbols of resistors, switches, fuses, batteries, inductors, capacitors, antennas, grounds and polarity; Construction of variable and fixed inductors and capacitors.

T7C01 (C)
Which of the following lists include three good electrical conductors?
A. Copper, gold, mica
B. Gold, silver, wood
C. Gold, silver, aluminum
D. Copper, aluminum, paper

T7C02 (D)
What is one reason resistors are used in electronic circuits?
A. To block the flow of direct current while allowing alternating current to pass

B. To block the flow of alternating current while allowing direct current to pass
C. To increase the voltage of the circuit
D. To control the amount of current that flows for a particular applied voltage

T7C03 (D)
If two resistors are connected in series, what is their total resistance?
A. The difference between the individual resistor values
B. Always less than the value of either resistor
C. The product of the individual resistor values
D. The sum of the individual resistor values

T7C04 (A)
What is one reason capacitors are used in electronic circuits?
A. To block the flow of direct current while allowing alternating current to pass
B. To block the flow of alternating current while allowing direct current to pass
C. To change the time constant of the applied voltage
D. To change alternating current to direct current

T7C05 (A)
If two equal-value capacitors are connected in parallel, what is their total capacitance?
A. Twice the value of one capacitor
B. Half the value of one capacitor
C. The same as the value of either capacitor
D. The value of one capacitor times the value of the other

T7C06 (B)
What does a capacitor do?
A. It stores energy electrochemically and opposes a change in current
B. It stores energy electrostatically and opposes a change in voltage
C. It stores energy electromagnetically and opposes a change in current
D. It stores energy electromechanically and opposes a change in voltage

T7C07 (D)
Which of the following best describes a variable capacitor?
A. A set of fixed capacitors whose connections can be varied
B. Two sets of insulating plates separated by a conductor, which can be varied in distance from each other
C. A set of capacitors connected in a series-parallel circuit
D. Two sets of rotating conducting plates separated by an insulator, which can be varied in surface area exposed to each other

T7C08 (C)
What does an inductor do?
A. It stores energy electrostatically and opposes a change in voltage
B. It stores energy electrochemically and opposes a change in current
C. It stores energy electromagnetically and opposes a change in current
D. It stores energy electromechanically and opposes a change in voltage

T7C09 (C)
What component controls current to flow in one direction only?
A. A fixed resistor
B. A signal generator
C. A diode
D. A fuse

T7C10 (A)
What is one advantage of using ICs (integrated circuits) instead of vacuum tubes in a circuit?
A. ICs usually combine several functions into one package
B. ICs can handle high-power input signals
C. ICs can handle much higher voltages
D. ICs can handle much higher temperatures

T7C11 (C)
Which symbol of Figure T7-1 represents a fixed resistor?
A. Symbol 1
B. Symbol 2

C. Symbol 3
D. Symbol 5

T7C12 (B)
In Figure T7-1, which symbol represents a variable resistor or potentiometer?
A. Symbol 1
B. Symbol 2
C. Symbol 3
D. Symbol 12

T7C13 (D)
In Figure T7-1, which symbol represents a single-cell battery?
A. Symbol 1
B. Symbol 6
C. Symbol 12
D. Symbol 13

T7C14 (B)
In Figure T7-1, which symbol represents an NPN transistor?
A. Symbol 2
B. Symbol 4
C. Symbol 10
D. Symbol 12

T7C15 (A)
Which symbol of Figure T7-1 represents a fixed-value capacitor?
A. Symbol 1
B. Symbol 3
C. Symbol 5
D. Symbol 13

T7C16 (B)
In Figure T7-1, which symbol represents an antenna?
A. Symbol 5
B. Symbol 7
C. Symbol 8
D. Symbol 14

T7C17 (A)
In Figure T7-1, which symbol represents a fixed-value iron-core inductor?
A. Symbol 6
B. Symbol 9
C. Symbol 11
D. Symbol 12

T7C18 (D)
In Figure T7-2, which symbol represents a single-pole, double-throw switch?
A. Symbol 1
B. Symbol 2
C. Symbol 3
D. Symbol 4

T7C19 (C)
In Figure T7-2, which symbol represents a double-pole, single-throw switch?
A. Symbol 1
B. Symbol 2
C. Symbol 3
D. Symbol 4

SUBELEMENT T8 - Good Engineering Practice [6 Exam Questions - 6 Groups]
T8A Basic amateur station apparatus; Choice of apparatus for desired communications; Setting up station; Constructing and modifying amateur station apparatus; Station layout for CW, SSB, FM, Packet and other popular modes.

T8A01 (C)
What two bands are most commonly used by "dual band" hand-held transceivers?
A. 6 meters and 2 meters
B. 2 meters and 1.25 meters
C. 2 meters and 70 cm
D. 70 cm and 23 cm

T8A02 (A)
If your mobile transceiver works in your car but not in your home, what should you check first?
A. The power supply
B. The speaker
C. The microphone
D. The SWR meter

T8A03 (B)
Which of the following devices would you need to conduct Amateur Radio communications

using a data emission?
A. A telegraph key
B. A computer
C. A transducer
D. A telemetry sensor

T8A04 (B)
Which of the following devices would be useful to create an effective Amateur Radio station for weak-signal VHF communication?
A. A hand-held VHF FM transceiver
B. A multi-mode VHF transceiver
C. An Omni-directional antenna
D. A mobile VHF FM transceiver

T8A05 (D)
What would you connect to a transceiver for voice operation?
A. A splatter filter
B. A terminal-voice controller
C. A receiver audio filter
D. A microphone

T8A06 (B)
What would you connect to a transceiver to send Morse code?
A. A key-click filter
B. A telegraph key
C. An SWR meter
D. An antenna switch

T8A07 (B)
What do many amateurs use to help form good Morse code characters?
A. A key-operated on/off switch
B. An electronic keyer
C. A key-click filter
D. A DTMF keypad

T8A08 (D)
Why is it important to provide adequate power supply filtering for a CW transmitter?
A. It isn't important, since CW transmitters cannot be modulated by AC hum
B. To eliminate phase noise
C. It isn't important, since most CW receivers can easily suppress any hum by using narrow filters
D. To eliminate modulation of the RF signal by AC hum

T8A09 (C)
Why is it important to provide adequate DC source supply filtering for a mobile transmitter or transceiver?
A. To reduce AC hum and carrier current device signals
B. To provide an emergency power source
C. To reduce stray noise and RF pick-up
D. To allow the use of smaller power conductors

T8A10 (A)
What would you connect to a transceiver for RTTY operation?
A. A modem and a teleprinter or computer system
B. A computer, a printer and a RTTY refresh unit
C. A data-inverter controller
D. A modem, a monitor and a DTMF keypad

T8A11 (B)
What might you connect between your transceiver and an antenna switch connected to several antennas?
A. A high-pass filter
B. An SWR meter
C. A key-click filter
D. A mixer

T8A12 (A)
What might happen if you set your receiver's signal squelch too low while attempting to receive packet mode transmissions?
A. Noise may cause the TNC to falsely detect a data carrier
B. Weaker stations may not be received
C. Transmission speed and throughput will be reduced
D. The TNC could be damaged

T8A13 (D)
What is one common method of transmitting RTTY on VHF/UHF bands?
A. Frequency shift the carrier to indicate mark and space at the receiver

B. Amplitude shift the carrier to indicate mark and space at the receiver
C. Key the transmitter on to indicate space and off for mark
D. Modulate a conventional FM transmitter with a modem

T8A14 (D)
What would you use to connect a dual-band antenna to a mobile transceiver that has separate VHF and UHF output connectors?
A. A dual-needle SWR meter
B. A full-duplex phone patch
C. Twin high-pass filters
D. A duplexer

T8B How transmitters work; Operation and tuning; VFO; Transceiver; Dummy load; Antenna switch; Power supply; Amplifier; Stability; Microphone gain; FM deviation; Block diagrams of typical stations.

T8B01 (B)
Can a transceiver designed for FM phone operation also be used for single sideband in the weak-signal portion of the 2-meter band?
A. Yes, with simple modification
B. Only if the radio is a "multimode" radio
C. Only with the right antenna
D. Only with the right polarization

T8B02 (B)
How is a CW signal usually transmitted?
A. By frequency-shift keying an RF signal
B. By on/off keying an RF signal
C. By audio-frequency-shift keying an oscillator tone
D. By on/off keying an audio-frequency signal

T8B03 (B)
What purpose does block 1 serve in the simple CW transmitter pictured in Figure T8-1?
A. It detects the CW signal
B. It controls the transmitter frequency
C. It controls the transmitter output power
D. It filters out spurious emissions from the transmitter

T8B04 (D)
What circuit is pictured in Figure T8-1 if block 1 is a variable-frequency oscillator?
A. A packet-radio transmitter
B. A crystal-controlled transmitter
C. A single-sideband transmitter
D. A VFO-controlled transmitter

T8B05 (C)
What circuit is shown in Figure T8-2 if block 1 represents a reactance modulator?
A. A single-sideband transmitter
B. A double-sideband AM transmitter
C. An FM transmitter
D. A product transmitter

T8B06 (D)
How would the output of the FM transmitter shown in Figure T8-2 be affected if the audio amplifier failed to operate (assuming block 1 is a reactance modulator)?
A. There would be no output from the transmitter
B. The output would be 6-dB below the normal output power
C. The transmitted audio would be distorted but understandable
D. The output would be an unmodulated carrier

T8B07 (A)
What minimum rating should a dummy antenna have for use with a 100-watt, single-sideband-phone transmitter?
A. 100 watts continuous
B. 141 watts continuous
C. 175 watts continuous
D. 200 watts continuous

T8B08 (B)
A mobile radio may be operated at home with the addition of which piece of equipment?
A. An alternator
B. A power supply
C. A linear amplifier
D. A rhombic antenna

T8B09 (C)
What might you use instead of a power supply for

home operation of a mobile radio?
A. A filter capacitor
B. An alternator
C. A 12-volt battery
D. A linear amplifier

T8B10 (C)
What device converts 120 V AC to 12 V DC?
A. A catalytic converter
B. A low-pass filter
C. A power supply
D. An RS-232 interface

T8B11 (B)
What device could boost the low-power output from your hand-held radio up to 100 watts?
A. A voltage divider
B. A power amplifier
C. A impedance network
D. A voltage regulator

T8B12 (B)
What is the result of over deviation in an FM transmitter?
A. Increased transmitter power
B. Out-of-channel emissions
C. Increased transmitter range
D. Poor carrier suppression

T8B13 (D)
What can you do if you are told your FM hand-held or mobile transceiver is over deviating?
A. Talk louder into the microphone
B. Let the transceiver cool off
C. Change to a higher power level
D. Talk farther away from the microphone

T8B14 (B)
In Figure T8-3, if block 1 is a transceiver and block 3 is a dummy antenna, what is block 2?
A. A terminal-node switch
B. An antenna switch
C. A telegraph key switch
D. A high-pass filter

T8B15 (D)
In Figure T8-3, if block 1 is a transceiver and block 2 is an antenna switch, what is block 3?
A. A terminal-node switch
B. An SWR meter
C. A telegraph key switch
D. A dummy antenna

T8B16 (B)
In Figure T8-4, if block 1 is a transceiver and block 2 is an SWR meter, what is block 3?
A. An antenna switch
B. An antenna tuner
C. A key-click filter
D. A terminal-node controller

T8B17 (C)
In Figure T8-4, if block 1 is a transceiver and block 3 is an antenna tuner, what is block 2?
A. A terminal-node switch
B. A dipole antenna
C. An SWR meter
D. A high-pass filter

T8B18 (D)
In Figure T8-4, if block 2 is an SWR meter and block 3 is an antenna tuner, what is block 1?
A. A terminal-node switch
B. A power supply
C. A telegraph key switch
D. A transceiver

T8C How receivers work, operation and tuning, including block diagrams; Super-heterodyne including Intermediate frequency; Reception; Demodulation or Detection; Sensitivity; Selectivity; Frequency standards; Squelch and audio gain (volume) control.

T8C01 (C)
What type of circuit does Figure T8-5 represent if block 1 is a product detector?
A. A simple phase modulation receiver
B. A simple FM receiver
C. A simple CW and SSB receiver
D. A double-conversion multiplier

T8C02 (D)
If Figure T8-5 is a diagram of a simple single-sideband receiver, what type of circuit should be shown in block 1?

A. A high pass filter
B. A ratio detector
C. A low pass filter
D. A product detector

T8C03 (D)
What circuit is pictured in Figure T8-6, if block 1 is a frequency discriminator?
A. A double-conversion receiver
B. A variable-frequency oscillator
C. A superheterodyne receiver
D. An FM receiver

T8C04 (A)
What is block 1 in the FM receiver shown in Figure T8-6?
A. A frequency discriminator
B. A product detector
C. A frequency-shift modulator
D. A phase inverter

T8C05 (B)
What would happen if block 1 failed to function in the FM receiver diagram shown in Figure T8-6?
A. The audio output would sound loud and distorted
B. There would be no audio output
C. There would be no effect
D. The receiver's power supply would be short-circuited

T8C06 (C)
What circuit function is found in all types of receivers?
A. An audio filter
B. A beat-frequency oscillator
C. A detector
D. An RF amplifier

T8C07 (C)
What is one accurate way to check the calibration of your receiver's tuning dial?
A. Monitor the BFO frequency of a second receiver
B. Tune to a popular amateur net frequency
C. Tune to one of the frequencies of station WWV or WWVH
D. Tune to another amateur station and ask what frequency the operator is using

T8C08 (B)
What circuit combines signals from an IF amplifier stage and a beat-frequency oscillator (BFO), to produce an audio signal?
A. An AGC circuit
B. A detector circuit
C. A power supply circuit
D. A VFO circuit

T8C09 (C)
Why is FM voice so effective for local VHF/UHF radio communications?
A. The carrier is not detectable
B. It is more resistant to distortion caused by reflected signals than the AM modes
C. It has audio that is less affected by interference from static-type electrical noise than the AM modes
D. Its RF carrier stays on frequency better than the AM modes

T8C10 (D)
Why do many radio receivers have several IF filters of different bandwidths that can be selected by the operator?
A. Because some frequency bands are wider than others
B. Because different bandwidths help increase the receiver sensitivity
C. Because different bandwidths improve S-meter readings
D. Because some emission types need a wider bandwidth than others to be received properly

T8C11 (C)
What is the function of a mixer in a superheterodyne receiver?
A. To cause all signals outside of a receiver's passband to interfere with one another
B. To cause all signals inside of a receiver's passband to reinforce one another
C. To shift the frequency of the received signal so that it can be processed by IF stages
D. To interface the receiver with an auxiliary device, such as a TNC

T8C12 (D)
What frequency or frequencies could the radio shown in Figure T8-7 receive?
A. 136.3 MHz
B. 157.7 MHz and 10.7 MHz
C. 10.7 MHz
D. 147.0 MHz and 168.4 MHz

T8C13 (C)
What type of receiver is shown in Figure T8-7?
A. Direct conversion
B. Superregenerative
C. Single-conversion superhetrodyne
D. Dual conversion superhetrodyne

T8C14 (B)
What emission mode could the receiver in Figure T8-7 detect?
A. AM
B. FM
C. Single sideband (SSB)
D. CW

T8C15 (C)
Where should the squelch be set for the proper operation of an FM receiver?
A. Low enough to hear constant background noise
B. Low enough to hear chattering background noise
C. At the point that just silences background noise
D. As far beyond the point of silence as the knob will turn

T8D How antennas work; Radiation principles; Basic construction; Half wave dipole length vs. frequency; Polarization; Directivity; ERP; Directional/non-directional antennas; Multiband antennas; Antenna gain; Resonant frequency; Loading coil; Electrical vs. physical length; Radiation pattern; Transmatch.

T8D01 (C)
Which of the following will improve the operation of a hand-held radio inside a vehicle?
A. Shielding around the battery pack
B. A good ground to the belt clip
C. An external antenna on the roof
D. An audio amplifier

T8D02 (B)
Which is true of "rubber duck" antennas for hand-held transceivers?
A. The shorter they are, the better they perform
B. They are much less efficient than a quarter-wavelength telescopic antenna
C. They offer the highest amount of gain possible for any hand-held transceiver antenna
D. They have a good long-distance communications range

T8D03 (B)
What would be the length, to the nearest inch, of a half-wavelength dipole antenna that is resonant at 147 MHz?
A. 19 inches
B. 37 inches
C. 55 inches
D. 74 inches

T8D04 (C)
How long should you make a half-wavelength dipole antenna for 223 MHz (measured to the nearest inch)?
A. 112 inches
B. 50 inches
C. 25 inches
D. 12 inches

T8D05 (C)
How long should you make a quarter-wavelength vertical antenna for 146 MHz (measured to the nearest inch)?
A. 112 inches
B. 50 inches
C. 19 inches
D. 12 inches

T8D06 (C)
How long should you make a quarter-wavelength vertical antenna for 440 MHz (measured to the nearest inch)?
A. 12 inches

B. 9 inches
C. 6 inches
D. 3 inches

T8D07 (B)
Which of the following factors has the greatest effect on the gain of a properly designed Yagi antenna?
A. The number of elements
B. Boom length
C. Element spacing
D. Element diameter

T8D08 (C)
Approximately how long is the driven element of a Yagi antenna?
A. 1/4 wavelength
B. 1/3 wavelength
C. 1/2 wavelength
D. 1 wavelength

T8D09 (D)
In Figure T8-8, what is the name of element 2 of the Yagi antenna?
A. Director
B. Reflector
C. Boom
D. Driven element

T8D10 (A)
In Figure T8-8, what is the name of element 3 of the Yagi antenna?
A. Director
B. Reflector
C. Boom
D. Driven element

T8D11 (B)
In Figure T8-8, what is the name of element 1 of the Yagi antenna?
A. Director
B. Reflector
C. Boom
D. Driven element

T8D12 (B)
What is a cubical quad antenna?
A. Four straight, parallel elements in line with each other, each approximately 1/2-electrical wavelength long
B. Two or more parallel four-sided wire loops, each approximately one-electrical wavelength long
C. A vertical conductor 1/4-electrical wavelength high, fed at the bottom
D. A center-fed wire 1/2-electrical wavelength long

T8D13 (B)
What does horizontal wave polarization mean?
A. The magnetic lines of force of a radio wave are parallel to the Earth's surface
B. The electric lines of force of a radio wave are parallel to the Earth's surface
C. The electric lines of force of a radio wave are perpendicular to the Earth's surface
D. The electric and magnetic lines of force of a radio wave are perpendicular to the Earth's surface

T8D14 (C)
What does vertical wave polarization mean?
A. The electric lines of force of a radio wave are parallel to the Earth's surface
B. The magnetic lines of force of a radio wave are perpendicular to the Earth's surface
C. The electric lines of force of a radio wave are perpendicular to the Earth's surface
D. The electric and magnetic lines of force of a radio wave are parallel to the Earth's surface

T8D15 (C)
If the ends of a half-wavelength dipole antenna (mounted at least a half-wavelength high) point east and west, which way would the antenna send out radio energy?
A. Equally in all directions
B. Mostly up and down
C. Mostly north and south
D. Mostly east and west

T8D16 (B)
What electromagnetic wave polarization do most repeater antennas have in the VHF and UHF spectrum?
A. Horizontal

B. Vertical
C. Right-hand circular
D. Left-hand circular

T8D17 (C)
What electromagnetic wave polarization is used for most satellite operation?
A. Only horizontal
B. Only vertical
C. Circular
D. No polarization

T8D18 (B)
Which antenna polarization is used most often for weak signal VHF/UHF SSB operation?
A. Vertical
B. Horizontal
C. Right-hand circular
D. Left-hand circular

T8D19 (C)
How will increasing antenna gain by 3 dB affect your signal's effective radiated power in the direction of maximum radiation?
A. It will cut it in half
B. It will not change
C. It will double it
D. It will quadruple it

T8D20 (A)
What is one advantage to using a multiband antenna?
A. You can operate on several bands with a single feed line
B. Multiband antennas always have high gain
C. You can transmit on several frequencies simultaneously
D. Multiband antennas offer poor harmonic suppression

T8D21 (D)
What could be done to reduce the physical length of an antenna without changing its resonant frequency?
A. Attach a balun at the feed point
B. Add series capacitance at the feed point
C. Use thinner conductors
D. Add a loading coil

T8D22 (C)
What device might allow use of an antenna on a band it was not designed for?
A. An SWR meter
B. A low-pass filter
C. An antenna tuner
D. A high-pass filter

T8E How transmission lines work; Standing waves/SWR/SWR-meter; Impedance matching; Types of transmission lines; Feed point; Coaxial cable; Balun; Waterproofing Connections.

T8E01 (D)
What does standing-wave ratio mean?
A. The ratio of maximum to minimum inductances on a feed line
B. The ratio of maximum to minimum capacitances on a feed line
C. The ratio of maximum to minimum impedances on a feed line
D. The ratio of maximum to minimum voltages on a feed line

T8E02 (C)
What instrument is used to measure standing wave ratio?
A. An ohmmeter
B. An ammeter
C. An SWR meter
D. A current bridge

T8E03 (D)
What would an SWR of 1:1 indicate about an antenna system?
A. That the antenna was very effective
B. That the transmission line was radiating
C. That the antenna was reflecting as much power as it was radiating
D. That the impedance of the antenna and its transmission line were matched

T8E04 (D)
What does an SWR reading of 4:1 mean?
A. An impedance match that is too low
B. An impedance match that is good, but not the best

C. An antenna gain of 4
D. An impedance mismatch; something may be wrong with the antenna system

T8E05 (A)
What does an antenna tuner do?
A. It matches a transceiver output impedance to the antenna system impedance
B. It helps a receiver automatically tune in stations that are far away
C. It switches an antenna system to a transceiver when sending, and to a receiver when listening
D. It switches a transceiver between different kinds of antennas connected to one feed line

T8E06 (D)
What is a coaxial cable?
A. Two wires side-by-side in a plastic ribbon
B. Two wires side-by-side held apart by insulating rods
C. Two wires twisted around each other in a spiral
D. A center wire inside an insulating material covered by a metal sleeve or shield

T8E07 (A)
Why should you use only good quality coaxial cable and connectors for a UHF antenna system?
A. To keep RF loss low
B. To keep television interference high
C. To keep the power going to your antenna system from getting too high
D. To keep the standing-wave ratio of your antenna system high

T8E08 (B)
What is parallel-conductor feed line?
A. Two wires twisted around each other in a spiral
B. Two wires side-by-side held apart by insulating material
C. A center wire inside an insulating material that is covered by a metal sleeve or shield
D. A metal pipe that is as wide or slightly wider than a wavelength of the signal it carries

T8E09 (D)
Which of the following are some reasons to use parallel-conductor, open-wire feed line?
A. It has low impedance and will operate with a high SWR
B. It will operate well even with a high SWR and it works well when tied down to metal objects
C. It has a low impedance and has less loss than coaxial cable
D. It will operate well even with a high SWR and has less loss than coaxial cable

T8E10 (D)
What does "balun" mean?
A. Balanced antenna network
B. Balanced unloader
C. Balanced unmodulator
D. Balanced to unbalanced

T8E11 (A)
Where would you install a balun to feed a dipole antenna with 50-ohm coaxial cable?
A. Between the coaxial cable and the antenna
B. Between the transmitter and the coaxial cable
C. Between the antenna and the ground
D. Between the coaxial cable and the ground

T8E12 (C)
What happens to radio energy when it is sent through a poor quality coaxial cable?
A. It causes spurious emissions
B. It is returned to the transmitter's chassis ground
C. It is converted to heat in the cable
D. It causes interference to other stations near the transmitting frequency

T8E13 (C)
What is an unbalanced line?
A. A feed line with neither conductor connected to ground
B. A feed line with both conductors connected to ground
C. A feed line with one conductor connected to ground
D. All of these answers are correct

T8E14 (C)
What point in an antenna system is called the feed point?
A. The antenna connection on the back of the transmitter
B. Halfway between the transmitter and the feed line
C. At the point where the feed line joins the antenna
D. At the tip of the antenna

T8F
Voltmeter/ammeter/ohmmeter/multi/S-meter, peak reading and RF watt meter; Building/modifying equipment; Soldering; Making measurements; Test instruments.

T8F01 (B)
Which instrument would you use to measure electric potential or electromotive force?
A. An ammeter
B. A voltmeter
C. A wavemeter
D. An ohmmeter

T8F02 (B)
How is a voltmeter usually connected to a circuit under test?
A. In series with the circuit
B. In parallel with the circuit
C. In quadrature with the circuit
D. In phase with the circuit

T8F03 (A)
What happens inside a voltmeter when you switch it from a lower to a higher voltage range?
A. Resistance is added in series with the meter
B. Resistance is added in parallel with the meter
C. Resistance is reduced in series with the meter
D. Resistance is reduced in parallel with the meter

T8F04 (A)
How is an ammeter usually connected to a circuit under test?
A. In series with the circuit
B. In parallel with the circuit
C. In quadrature with the circuit
D. In phase with the circuit

T8F05 (D)
Which instrument would you use to measure electric current?
A. An ohmmeter
B. A wavemeter
C. A voltmeter
D. An ammeter

T8F06 (D)
What test instrument would be useful to measure DC resistance?
A. An oscilloscope
B. A spectrum analyzer
C. A noise bridge
D. An ohmmeter

T8F07 (C)
What might damage a multimeter that uses a moving-needle meter?
A. Measuring a voltage much smaller than the maximum for the chosen scale
B. Leaving the meter in the milliamps position overnight
C. Measuring voltage when using the ohms setting
D. Not allowing it to warm up properly

T8F08 (D)
For which of the following measurements would you normally use a multimeter?
A. SWR and power
B. Resistance, capacitance and inductance
C. Resistance and reactance
D. Voltage, current and resistance

T8F09 (A)
What is used to measure relative signal strength in a receiver?
A. An S meter
B. An RST meter
C. A signal deviation meter
D. An SSB meter

T8F10 (A)
With regard to a transmitter and antenna system, what does "forward power" mean?

A. The power traveling from the transmitter to the antenna
B. The power radiated from the top of an antenna system
C. The power produced during the positive half of an RF cycle
D. The power used to drive a linear amplifier

T8F11 (B)
With regard to a transmitter and antenna system, what does "reflected power" mean?
A. The power radiated down to the ground from an antenna
B. The power returned towards the source on a transmission line
C. The power produced during the negative half of an RF cycle
D. The power returned to an antenna by buildings and trees

T8F12 (B)
At what line impedance do most RF watt meters usually operate?
A. 25 ohms
B. 50 ohms
C. 100 ohms
D. 300 ohms

T8F13 (B)
If a directional RF wattmeter reads 90 watts forward power and 10 watts reflected power, what is the actual transmitter output power?
A. 10 watts
B. 80 watts
C. 90 watts
D. 100 watts

T8F14 (B)
What is the minimum FCC certification required for an Amateur Radio operator to build or modify their own transmitting equipment?
A. A First-Class Radio Repair License
B. A Technician class license
C. A General class license
D. An Amateur Extra class license

T8F15 (D)
What safety step should you take when soldering?
A. Always wear safety glasses
B. Ensure proper ventilation
C. Make sure no one can touch the soldering iron tip for at least 10 minutes after it is turned off
D. All of these choices are correct

T8F16 (D)
Where would you connect a voltmeter to a 12-volt transceiver if you think the supply voltage may be low when you transmit?
A. At the battery terminals
B. At the fuse block
C. Midway along the 12-volt power supply wire
D. At the 12-volt plug on the chassis of the equipment

T8F17 (D)
If your mobile transceiver does not power up, what might you check first?
A. The antenna feedpoint
B. The coaxial cable connector
C. The microphone jack
D. The 12-volt fuses

T8F18 (C)
What device produces a stable, low-level signal that can be set to a desired frequency?
A. A wavemeter
B. A reflectometer
C. A signal generator
D. An oscilloscope

T8F19 (D)
In Figure T8-9, what circuit quantity would meter B indicate?
A. The voltage across the resistor
B. The power consumed by the resistor
C. The power factor of the resistor
D. The current flowing through the resistor

T8F20 (B)
In Figure T8-9, what circuit quantity is meter A reading?
A. Battery current
B. Battery voltage
C. Battery power
D. Battery current polarity

T8F21 (A)
In Figure T8-9, how would the power consumed by the resistor be calculated?
A. Multiply the value of the resistor times the square of the reading of meter B
B. Multiply the value of the resistor times the reading of meter B
C. Multiply the reading of meter A times the value of the resistor
D. Multiply the value of the resistor times the square root of the reading of meter B

SUBELEMENT T9 - Special Operations [2 Exam Questions — 2 Groups]
T9A How an FM Repeater Works; Repeater operating procedures; Available frequencies; Input/output frequency separation; Repeater ID requirements; Simplex operation; Coordination; Time out; Open/closed repeater; Responsibility for interference.

T9A01 (B)
What is the purpose of repeater operation?
A. To cut your power bill by using someone else's higher power system
B. To help mobile and low-power stations extend their usable range
C. To transmit signals for observing propagation and reception
D. To communicate with stations in services other than amateur

T9A02 (B)
What is a courtesy tone, as used in repeater operations?
A. A sound used to identify the repeater
B. A sound used to indicate when a transmission is complete
C. A sound used to indicate that a message is waiting for someone
D. A sound used to activate a receiver in case of severe weather

T9A03 (D)
During commuting rush hours, which type of repeater operation should be discouraged?
A. Mobile stations
B. Low-power stations
C. Highway traffic information nets
D. Third-party communications nets

T9A04 (D)
Which of the following is a proper way to break into a conversation on a repeater?
A. Wait for the end of a transmission and start calling the desired party
B. Shout, "break, break!" to show that you're eager to join the conversation
C. Turn on an amplifier and override whoever is talking
D. Say your call sign during a break between transmissions

T9A05 (A)
When using a repeater to communicate, which of the following do you need to know about the repeater?
A. Its input frequency and offset
B. Its call sign
C. Its power level
D. Whether or not it has an autopatch

T9A06 (C)
Why should you pause briefly between transmissions when using a repeater?
A. To check the SWR of the repeater
B. To reach for pencil and paper for third-party communications
C. To listen for anyone wanting to break in
D. To dial up the repeater's autopatch

T9A07 (A)
Why should you keep transmissions short when using a repeater?
A. A long transmission may prevent someone with an emergency from using the repeater
B. To see if the receiving station operator is still awake
C. To give any listening non-hams a chance to respond
D. To keep long-distance charges down

T9A08 (A)
How could you determine if a repeater is already being used by other stations?

A. Ask if the frequency is in use, then give your call sign
B. If you don't hear anyone, assume that the frequency is clear to use
C. Check for the presence of the CTCSS tone
D. If the repeater identifies when you key your transmitter, it probably was already in use

T9A09 (A)
What is the usual input/output frequency separation for repeaters in the 2-meter band?
A. 600 kHz
B. 1.0 MHz
C. 1.6 MHz
D. 5.0 MHz

T9A10 (D)
What is the usual input/output frequency separation for repeaters in the 70-centimeter band?
A. 600 kHz
B. 1.0 MHz
C. 1.6 MHz
D. 5.0 MHz

T9A11 (A)
What does it mean to say that a repeater has an input and an output frequency?
A. The repeater receives on one frequency and transmits on another
B. The repeater offers a choice of operating frequency, in case one is busy
C. One frequency is used to control the repeater and another is used to retransmit received signals
D. The repeater must receive an access code on one frequency before retransmitting received signals

T9A12 (C)
What is the most likely reason you might hear Morse code tones on a repeater frequency?
A. Intermodulation
B. An emergency request for help
C. The repeater's identification
D. A courtesy tone

T9A13 (A)
What is the common amateur meaning of the term "simplex operation"?
A. Transmitting and receiving on the same frequency
B. Transmitting and receiving over a wide area
C. Transmitting on one frequency and receiving on another
D. Transmitting one-way communications

T9A14 (B)
When should you use simplex operation instead of a repeater?
A. When the most reliable communications are needed
B. When a contact is possible without using a repeater
C. When an emergency telephone call is needed
D. When you are traveling and need some local information

T9A15 (A)
If you are talking to a station using a repeater, how would you find out if you could communicate using simplex instead?
A. See if you can clearly receive the station on the repeater's input frequency
B. See if you can clearly receive the station on a lower frequency band
C. See if you can clearly receive a more distant repeater
D. See if a third station can clearly receive both of you

T9A16 (D)
What is it called if the frequency coordinator recommends that you operate on a specific repeater frequency pair?
A. FCC type acceptance
B. FCC type approval
C. Frequency division multiplexing
D. Repeater frequency coordination

T9A17 (D)
What is the purpose of a repeater time-out timer?
A. It lets a repeater have a rest period after heavy use
B. It logs repeater transmit time to predict when a repeater will fail
C. It tells how long someone has been using a repeater

D. It limits the amount of time a repeater can transmit continuously

T9A18 (A)
What should you do if you hear a closed repeater system that you would like to be able to use?
A. Contact the control operator and ask to join
B. Use the repeater until told not to
C. Use simplex on the repeater input until told not to
D. Write the FCC and report the closed condition

T9A19 (B)
Who pays for the site rental and upkeep of most repeaters?
A. All amateurs, because part of the amateur license examination fee is used
B. The repeater owner and donations from its users
C. The Federal Communications Commission
D. The federal government, using money granted by Congress

T9A20 (D) [97.205c]
If a repeater is causing harmful interference to another amateur repeater and a frequency coordinator has recommended the operation of both repeaters, who is responsible for resolving the interference?
A. The licensee of the repeater that has been recommended for the longest period of time
B. The licensee of the repeater that has been recommended the most recently
C. The frequency coordinator
D. Both repeater licensees

T9B Beacon, satellite, space, EME communications; Radio control of models; Autopatch; Slow scan television; Telecommand; CTCSS tone access; Duplex/crossband operation.

T9B01 (A) [97.3a9]
What is an amateur station called that transmits communications for the purpose of observation of propagation and reception?
A. A beacon
B. A repeater
C. An auxiliary station
D. A radio control station

T9B02 (D) [97.203c,d,g]
Which of the following is true of Amateur Radio beacon stations?
A. Automatic control is allowed in certain band segments
B. One-way transmissions are permitted
C. Maximum output power is 100 watts
D. All of these choices are correct

T9B03 (C) [97.209a]
The control operator of a station communicating through an amateur satellite must hold what class of license?
A. Amateur Extra or Advanced
B. Any class except Novice
C. Any class
D. Technician with satellite endorsement

T9B04 (B)
How does the Doppler effect change an amateur satellite's signal as the satellite passes overhead?
A. The signal's amplitude increases or decreases
B. The signal's frequency increases or decreases
C. The signal's polarization changes from horizontal to vertical
D. The signal's circular polarization rotates

T9B05 (D)
Why do many amateur satellites operate on the VHF/UHF bands?
A. To take advantage of the skip zone
B. Because VHF/UHF equipment costs less than HF equipment
C. To give Technician class operators greater access to modern communications technology
D. Because VHF and UHF signals easily pass through the ionosphere

T9B06 (C)
Which antenna system would NOT be a good choice for an EME (moonbounce) station?
A. A parabolic-dish antenna
B. A multi-element array of collinear antennas

C. A ground-plane antenna
D. A high-gain array of Yagi antennas

T9B07 (B)
What does the term "apogee" refer to when applied to an Earth satellite?
A. The closest point to the Earth in the satellite's orbit
B. The most distant point from the Earth in the satellite's orbit
C. The point where the satellite appears to cross the equator
D. The point when the Earth eclipses the satellite from the sun

T9B08 (A)
What does the term "perigee" refer to when applied to an Earth satellite?
A. The closest point to the Earth in the satellite's orbit
B. The most distant point from the Earth in the satellite's orbit
C. The time when the satellite will be on the opposite side of the Earth
D. The effect that causes the satellite's signal frequency to change

T9B09 (D)
What mathematical parameters describe a satellite's orbit?
A. Its telemetry data
B. Its Doppler shift characteristics
C. Its mean motion
D. Its Keplerian elements

T9B10 (A)
What is the typical amount of time an amateur has to communicate with the International Space Station?
A. 4 to 6 minutes per pass
B. An hour or two per pass
C. About 20 minutes per pass
D. All day

T9B11 (A)
Which of the following would be the best emission mode for two-way EME contacts?
A. CW
B. AM
C. FM
D. Spread spectrum

T9B12 (C) [97.215a]
What minimum information must be on a label affixed to a transmitter used for telecommand (control) of model craft?
A. Station call sign
B. Station call sign and the station licensee's name
C. Station call sign and the station licensee's name and address
D. Station call sign and the station licensee's class of license

T9B13 (C)
What is an autopatch?
A. An automatic digital connection between a US and a foreign amateur
B. A digital connection used to transfer data between a hand-held radio and a computer
C. A device that allows radio users to access the public telephone system
D. A video interface allowing images to be patched into a digital data stream

T9B14 (C)
Which of the following statements about Amateur Radio autopatch usage is true?
A. The person called using the autopatch must be a licensed radio amateur
B. The autopatch will allow only local calls to police, fire and ambulance services
C. Communication through the autopatch is not private
D. The autopatch should not be used for reporting emergencies

T9B15 (B)
Which of the following will allow you to monitor Amateur Television (ATV) on the 70-cm band?
A. A portable video camera
B. A cable ready TV receiver
C. An SSTV converter
D. A TV flyback transformer

T9B16 (A)
When may slow-scan television be transmitted through a 2-meter repeater?
A. At any time, providing the repeater control operator authorizes this unique transmission
B. Never; slow-scan television is not allowed on 2 meters
C. Only after 5:00 PM local time
D. Never; slow-scan television is not allowed on repeaters

T9B17 (C) [97.3a43]
What is the definition of telecommand?
A. All communications using the telephone or telegraphy with space stations
B. A one way transmission to initiate conversation with astronauts aboard a satellite or space station
C. A one way transmission to initiate, modify or terminate functions of a device at a distance
D. Two way transmissions to initiate, modify or terminate functions of a device at a distance

T9B18 (D) [97.213a,b,c]
What provisions must be in place for the legal operation of a telecommand station?
A. The station must have a wire line or radio control link
B. A photocopy of the station license must be posted in a conspicuous location
C. The station must be protected so no unauthorized transmission can be made
D. All of these choices are correct

T9B19 (B)
What is a continuous tone-coded squelch system (CTCSS) tone (sometimes called PL — a Motorola trademark)?
A. A special signal used for telecommand control of model craft
B. A sub-audible tone, added to a carrier, which may cause a receiver to accept the signal
C. A tone used by repeaters to mark the end of a transmission
D. A special signal used for telemetry between amateur space stations and Earth stations

T9B20 (D)
What does it mean if you are told that a tone is required to access a repeater?
A. You must use keypad tones like your phone system to operate it
B. You must wait to hear a warbling two-tone signal to operate it
C. You must wait to hear a courtesy beep tone at the end of another's transmission before you can operate it
D. You must use a subaudible tone-coded squelch with your signal to operate it

T9B21 (D)
What is the term that describes a repeater that receives signals on one band and retransmits them on another band?
A. A special coordinated repeater
B. An illegally operating repeater
C. An auxiliary station
D. A crossband repeater

SUBELEMENT T0 - Electrical, Antenna Structure and RF Safety Practices [6 Exam Questions - 6 Groups]
T0A Sources of electrical danger in amateur stations: lethal voltages, high current sources, fire; avoiding electrical shock; Station wiring; Wiring a three wire electrical plug; Need for main power switch; Safety interlock switch; Open/short circuit; Fuses; Station grounding.

T0A01 (A)
What is the minimum voltage that is usually dangerous to humans?
A. 30 volts
B. 100 volts
C. 1000 volts
D. 2000 volts

T0A02 (D)
Which electrical circuit draws high current?
A. An open circuit
B. A dead circuit
C. A closed circuit
D. A short circuit

T0A03 (C)
What could happen to your transceiver if you

replace its blown 5 amp AC line
fuse with a 30 amp fuse?
A. The 30-amp fuse would better protect your transceiver from using too much current
B. The transceiver would run cooler
C. The transceiver could use more current than 5 amps and a fire could occur
D. The transceiver would not be able to produce as much RF output

T0A04 (A)
How much electrical current flowing through the human body will probably be fatal?
A. As little as 1/10 of an ampere
B. Approximately 10 amperes
C. More than 20 amperes
D. Current through the human body is never fatal

T0A05 (A)
Which body organ can be fatally affected by a very small amount of electrical current?
A. The heart
B. The brain
C. The liver
D. The lungs

T0A06 (B)
For best protection from electrical shock, what should be grounded in an amateur station?
A. The power supply primary
B. All station equipment connected to a common ground
C. The antenna feed line
D. The AC power mains

T0A07 (D)
Which potential does the green wire in a three-wire electrical plug represent?
A. Neutral
B. Hot
C. Hot and neutral
D. Ground

T0A08 (C)
What is an important consideration for the location of the main power switch?
A. It must always be near the operator
B. It must always be as far away from the operator as possible
C. Everyone should know where it is located in case of an emergency
D. It should be located in a locked metal box so no one can accidentally turnit off

T0A09 (A)
What circuit should be controlled by a safety interlock switch in an amateur transceiver or power amplifier?
A. The power supply
B. The IF amplifier
C. The audio amplifier
D. The cathode bypass circuit

T0A10 (C)
What type of electrical circuit is created when a fuse blows?
A. A closed circuit
B. A bypass circuit
C. An open circuit
D. A short circuit

T0A11 (D)
Why would it be unwise to touch an ungrounded terminal of a high voltage capacitor even if it's not in an energized circuit?
A. You could damage the capacitor's dielectric material
B. A residual charge on the capacitor could cause interference to others
C. You could damage the capacitor by causing an electrostatic discharge
D. You could receive a shock from a residual stored charge

T0A12 (A)
What safety equipment item should you always add to home built equipment that is powered from 110 volt AC lines?
A. A fuse or circuit breaker in series with the equipment
B. A fuse or circuit breaker in parallel with the equipment
C. Install Zener diodes across AC inputs
D. House the equipment in a plastic or other non-conductive enclosure

T0A13 (D)
When fuses are installed in 12-volt DC wiring, where should they be placed?
A. At the radio
B. Midway between voltage source and radio
C. Fuses aren't required for 12-volt DC equipment
D. At the voltage source

T0B Lightning protection; Antenna structure installation safety; Tower climbing Safety; Safety belt/hard hat/safety glasses; Antenna structure limitations.

T0B01 (C)
How can an antenna system best be protected from lightning damage?
A. Install a balun at the antenna feed point
B. Install an RF choke in the antenna feed line
C. Ground all antennas when they are not in use
D. Install a fuse in the antenna feed line

T0B02 (D)
How can amateur station equipment best be protected from lightning damage?
A. Use heavy insulation on the wiring
B. Never turn off the equipment
C. Disconnect the ground system from all radios
D. Disconnect all equipment from the power lines and antenna cables

T0B03 (C)
Why should you wear a hard hat and safety glasses if you are on the ground helping someone work on an antenna tower?
A. So you won't be hurt if the tower should accidentally fall
B. To keep RF energy away from your head during antenna testing
C. To protect your head from something dropped from the tower
D. So someone passing by will know that work is being done on the tower and will stay away

T0B04 (D)
What safety factors must you consider when using a bow and arrow or slingshot and weight to shoot an antenna-support line over a tree?
A. You must ensure that the line is strong enough to withstand the shock of shooting the weight
B. You must ensure that the arrow or weight has a safe flight path if the line breaks
C. You must ensure that the bow and arrow or slingshot is in good working condition
D. All of these choices are correct

T0B05 (B)
Which of the following is the best way to install your antenna in relation to overhead electric power lines?
A. Always be sure your antenna wire is higher than the power line, and crosses it at a 90-degree angle
B. Always be sure your antenna and feed line are well clear of any power lines
C. Always be sure your antenna is lower than the power line, and crosses it at a small angle
D. Only use vertical antennas within 100 feet of a power line

T0B06 (C)
What should you always do before attempting to climb an antenna tower?
A. Turn on all radio transmitters that use the tower's antennas
B. Remove all tower grounding to guard against static electric shock
C. Put on your safety belt and safety glasses
D. Inform the FAA and the FCC that you are starting work on a tower

T0B07 (D)
What is the most important safety precaution to take when putting up an antenna tower?
A. Install steps on your tower for safe climbing
B. Insulate the base of the tower to avoid lightning strikes
C. Ground the base of the tower to avoid lightning strikes
D. Look for and stay clear of any overhead electrical wires

T0B08 (A)
What should you consider before you climb a tower with a leather climbing belt?

A. If the leather is old, it is probably brittle and could break unexpectedly
B. If the leather is old, it is very tough and is not likely to break easily
C. If the leather is old, it is flexible and will hold you more comfortably
D. An unbroken old leather belt has proven its holding strength over the years

T0B09 (D)
What should you do before you climb a guyed tower?
A. Tell someone that you will be up on the tower
B. Inspect the tower for cracks or loose bolts
C. Inspect the guy wires for frayed cable, loose cable clamps, loose turnbuckles or loose guy anchors
D. All of these choices are correct

T0B10 (D)
What should you do before you do any work on top of your tower?
A. Tell someone that you will be up on the tower
B. Bring a variety of tools with you to minimize your trips up and down the tower
C. Inspect the tower before climbing to become aware of any antennas or other obstacles that you may need to step around
D. All of these choices are correct

T0C Definition of RF radiation; Procedures for RF environmental safety; Definitions and guidelines.

T0C01 (A)
What is radio frequency radiation?
A. Waves of electric and magnetic energy between 3 kHz and 300 GHz
B. Ultra-violet rays emitted by the sun between 20 Hz and 300 GHz
C. Sound energy given off by a radio receiver
D. Beams of X-Rays and Gamma rays emitted by a radio transmitter

T0C02 (B)
Why is it a good idea to adhere to the FCC's Rules for using the minimum power needed when you are transmitting with your hand-held radio?
A. Large fines are always imposed on operators violating this rule
B. To reduce the level of RF radiation exposure to the operator's head
C. To reduce calcification of the NiCd battery pack
D. To eliminate self-oscillation in the receiver RF amplifier

T0C03 (A)
Which of the following units of measurement are used to specify the power density of a radiated RF signal?
A. Milliwatts per square centimeter
B. Volts per meter
C. Amperes per meter
D. All of these choices are correct

T0C04 (D)
Over what frequency range are the FCC Regulations most stringent for RF radiation exposure?
A. Frequencies below 300 kHz
B. Frequencies between 300 kHz and 3 MHz
C. Frequencies between 3 MHz and 30 MHz
D. Frequencies between 30 MHz and 300 MHz

T0C05 (B)
Which of the following categories describes most common amateur use of a hand-held transceiver?
A. Mobile devices
B. Portable devices
C. Fixed devices
D. None of these choices is correct

T0C06 (D)
From an RF safety standpoint, what impact does the duty cycle have on the minimum safe distance separating an antenna and people in the neighboring environment?
A. The lower the duty cycle, the shorter the compliance distance
B. The compliance distance is increased with an increase in the duty cycle
C. Lower duty cycles subject people in the

environment to lower radio-frequency radiation
D. All of these answers are correct

T0C07 (A)
Why is the concept of "duty cycle" one factor used to determine safe RF radiation exposure levels?
A. It takes into account the amount of time the transmitter is operating at full power during a single transmission
B. It takes into account the transmitter power supply rating
C. It takes into account the antenna feed line loss
D. It takes into account the thermal effects of the final amplifier

T0C08 (D)
What factors affect the resulting RF fields emitted by an amateur transceiver
that expose people in the environment?
A. Frequency and power level of the RF field
B. Antenna height and distance from the antenna to a person
C. Radiation pattern of the antenna
D. All of these answers are correct

T0C09 (B)
What unit of measurement specifies RF electric field strength?
A. Coulombs (C) at one wavelength from the antenna
B. Volts per meter (V/m)
C. Microfarads (uF) at the transmitter output
D. Microhenrys (uH) per square centimeter

T0C10 (D)
Which of the following is considered to be non-Ionizing radiation?
A. X-radiation
B. Gamma radiation
C. Ultra violet radiation
D. Radio frequency radiation

T0C11 (C)
What do the FCC RF radiation exposure regulations establish?
A. Maximum radiated field strength
B. Minimum permissible HF antenna height
C. Maximum permissible exposure limits
D. All of these choices are correct

T0C12 (C)
Which of the following steps would help you to comply with RF-radiation exposure guidelines for uncontrolled RF environments?
A. Reduce transmitting times within a 6-minute period to reduce the station duty cycle
B. Operate only during periods of high solar absorption
C. Reduce transmitting times within a 30-minute period to reduce the station duty cycle
D. Operate only on high duty cycle modes

T0C13 (C)
Which of the following steps would help you to comply with RF-exposure guidelines for controlled RF environments?
A. Reduce transmitting times within a 30-minute period to reduce the station duty cycle
B. Operate only during periods of high solar absorption
C. Reduce transmitting times within a 6-minute period to reduce the station duty cycle
D. Operate only on high duty cycle modes

T0C14 (B)
To avoid excessively high human exposure to RF fields, how should amateur antennas generally be mounted?
A. With a high current point near ground
B. As far away from accessible areas as possible
C. On a nonmetallic mast
D. With the elements in a horizontal polarization

T0C15 (D)
What action can amateur operators take to prevent exposure to RF radiation in
excess of the FCC-specified limits?
A. Alter antenna patterns
B. Relocate antennas
C. Revise station technical parameters, such as

frequency, power, or emission type
D. All of these choices are correct

T0C16 (C)
Which of the following radio frequency emissions will result in the least RF radiation exposure if they all have the same peak envelope power (PEP)?
A. Two-way exchanges of phase-modulated (PM) telephony
B. Two-way exchanges of frequency-modulated (FM) telephony
C. Two-way exchanges of single-sideband (SSB) telephony
D. Two-way exchanges of Morse code (CW) communication

T0C17 (C)
Why is the concept of "specific absorption rate (SAR)" one factor used to determine safe RF radiation exposure levels?
A. It takes into account the overall efficiency of the final amplifier
B. It takes into account the transmit/receive time ratio during normal amateur communication
C. It takes into account the rate at which the human body absorbs RF energy at a particular frequency
D. It takes into account the antenna feed line loss

T0C18 (D)
Why must the frequency of an RF source be considered when evaluating RF radiation exposure?
A. Lower-frequency RF fields have more energy than higher-frequency fields
B. Lower-frequency RF fields penetrate deeper into the body than higher-frequency fields
C. Higher-frequency RF fields are transient in nature, and do not affect the human body
D. The human body absorbs more RF energy at some frequencies than at others

T0C19 (C)
What is the maximum power density that may be emitted from an amateur station under the FCC RF radiation exposure limits?
A. The FCC Rules specify a maximum emission of 1.0 milliwatt per square centimeter
B. The FCC Rules specify a maximum emission of 5.0 milliwatts per square centimeter
C. The FCC Rules specify exposure limits, not emission limits
D. The FCC Rules specify maximum emission limits that vary with frequency

T0D Radiofrequency exposure standards; Near/far field, Field strength; Compliance distance; Controlled/Uncontrolled environment.

T0D01 (A)
What factors must you consider if your repeater station antenna will be located at a site that is occupied by antennas for transmitters in other services?
A. Your radiated signal must be considered as part of the total RF radiation from the site when determining RF radiation exposure levels
B. Each individual transmitting station at a multiple transmitter site must
meet the RF radiation exposure levels
C. Each station at a multiple-transmitter site may add no more than 1% of the maximum permissible exposure (MPE) for that site
D. Amateur stations are categorically excluded from RF radiation exposure evaluation at multiple-transmitter sites

T0D02 (C)
Why do exposure limits vary with frequency?
A. Lower-frequency RF fields have more energy than higher-frequency fields
B. Lower-frequency RF fields penetrate deeper into the body than higher-frequency fields
C. The body's ability to absorb RF energy varies with frequency
D. It is impossible to measure specific absorption rates at some frequencies

T0D03 (C)
Why might mobile transceivers produce less RF radiation exposure than hand-held transceivers in mobile operations?

A. They do not produce less exposure because they usually have higher power levels.
B. They have a higher duty cycle
C. When mounted on a metal vehicle roof, mobile antennas are generally well shielded from vehicle occupants
D. Larger transmitters dissipate heat and energy more readily

T0D04 (C)
In the far field, as the distance from the source increases, how does power density vary?
A. The power density is proportional to the square of the distance
B. The power density is proportional to the square root of the distance
C. The power density is proportional to the inverse square of the distance
D. The power density is proportional to the inverse cube of the distance

T0D05 (D)
In the near field, how does the field strength vary with distance from the source?
A. It always increases with the cube of the distance
B. It always decreases with the cube of the distance
C. It varies as a sine wave with distance
D. It depends on the type of antenna being used

T0D06 (A)
Why should you never look into the open end of a microwave feed horn antenna while the transmitter is operating?
A. You may be exposing your eyes to more than the maximum permissible exposure of RF radiation
B. You may be exposing your eyes to more than the maximum permissible exposure level of infrared radiation
C. You may be exposing your eyes to more than the maximum permissible exposure level of ultraviolet radiation
D. All of these choices are correct

T0D07 (A)
What factors determine the location of the boundary between the near and far fields of an antenna?
A. Wavelength and the physical size of the antenna
B. Antenna height and element length
C. Boom length and element diameter
D. Transmitter power and antenna gain

T0D08 (A)
Referring to Figure T0-1, which of the following equations should you use to calculate the maximum permissible exposure (MPE) on the Technician (with code credit) HF bands for a controlled RF radiation exposure environment?
A. Maximum permissible power density in mw per square cm equals 900 divided by the square of the operating frequency, in MHz
B. Maximum permissible power density in mw per square cm equals 180 divided by the square of the operating frequency, in MHz
C. Maximum permissible power density in mw per square cm equals 900 divided by the operating frequency, in MHz
D. Maximum permissible power density in mw per square cm equals 180 divided by the operating frequency, in MHz

T0D09 (B)
Referring to Figure T0-1, what is the formula for calculating the maximum permissible exposure (MPE) limit for uncontrolled environments on the 2-meter (146 MHz) band?
A. There is no formula, MPE is a fixed power density of 1.0 milliwatt per square centimeter averaged over any 6 minutes
B. There is no formula, MPE is a fixed power density of 0.2 milliwatt per square centimeter averaged over any 30 minutes
C. The MPE in milliwatts per square centimeter equals the frequency in megahertz divided by 300 averaged over any 6 minutes
D. The MPE in milliwatts per square centimeter equals the frequency in megahertz divided by 1500 averaged over any 30 minutes

T0D10 (A)
What is the minimum safe distance for a controlled RF radiation environment from a

station using a half-wavelength dipole antenna on 7 MHz at 100 watts PEP, as specified in Figure T0-2?
A. 1.4 foot
B. 2 feet
C. 3.1 feet
D. 6.5 feet

T0D11 (C)
What is the minimum safe distance for an uncontrolled RF radiation environment from a station using a 3-element "triband" Yagi antenna on 28 MHz at 100 watts PEP, as specified in Figure T0-2?
A. 7 feet
B. 11 feet
C. 24.5 feet
D. 34 feet

T0D12 (A)
What is the minimum safe distance for a controlled RF radiation environment from a station using a 146 MHz quarter-wave whip antenna at 10 watts, as specified in Figure T0-2?
A. 1.7 feet
B. 2.5 feet
C. 1.2 feet
D. 2 feet

T0D13 (A)
What is the minimum safe distance for a controlled RF radiation environment from a station using a 17-element Yagi on a five-wavelength boom on 144 MHz at 100 watts, as specified in Figure T0-2?
A. 72.4 feet
B. 78.5 feet
C. 101 feet
D. 32.4 feet

T0D14 (B)
What is the minimum safe distance for an uncontrolled RF radiation environment from a station using a 446 MHz 5/8-wave ground plane vertical antenna at 10 watts, as specified in Figure T0-2?
A. 1 foot
B. 4.3 feet
C. 9.6 feet
D. 6 feet

T0E RF Biological effects and potential hazards; Radiation exposure limits; OET Bulletin 65; MPE (Maximum permissible exposure).

T0E01 (A)
If you do not have the equipment to measure the RF power densities present at your station, what might you do to ensure compliance with the FCC RF radiation exposure limits?
A. Use one or more of the methods included in the amateur supplement to FCC OET Bulletin 65
B. Call an FCC-Certified Test Technician to perform the measurements for you
C. Reduce power from 200 watts PEP to 100 watts PEP
D. Operate only low-duty-cycle modes such as FM

T0E02 (C)
Where will you find the applicable FCC RF radiation maximum permissible exposure (MPE) limits defined?
A. FCC Part 97 Amateur Service Rules and Regulations
B. FCC Part 15 Radiation Exposure Rules and Regulations
C. FCC Part 1 and Office of Engineering and Technology (OET) Bulletin 65
D. Environmental Protection Agency Regulation 65

T0E03 (D)
To determine compliance with the maximum permitted exposure (MPE) levels, safe exposure levels for RF energy are averaged for an "uncontrolled" RF environment over what time period?
A. 6 minutes
B. 10 minutes
C. 15 minutes
D. 30 minutes

T0E04 (A)
To determine compliance with the maximum

permitted exposure (MPE) levels, safe exposure levels for RF energy are averaged for a "controlled" RF environment over what time period?
A. 6 minutes
B. 10 minutes
C. 15 minutes
D. 30 minutes

T0E05 (D)
Why are Amateur Radio operators required to meet the FCC RF radiation exposure limits?
A. The standards are applied equally to all radio services
B. To ensure that RF radiation occurs only in a desired direction
C. Because amateur station operations are more easily adjusted than those of commercial radio services
D. To ensure a safe operating environment for amateurs, their families and neighbors

T0E06 (C)
At what frequencies do the FCC's RF radiation exposure guidelines incorporate limits for Maximum Permissible Exposure (MPE)?
A. All frequencies below 30 MHz
B. All frequencies between 20,000 Hz and 10 MHz
C. All frequencies between 300 kHz and 100 GHz
D. All frequencies above 300 GHz

T0E07 (D)
On what value are the maximum permissible exposure (MPE) limits based?
A. The square of the mass of the exposed body
B. The square root of the mass of the exposed body
C. The whole-body specific gravity (WBSG)
D. The whole-body specific absorption rate (SAR)

T0E08 (C)
What is one biological effect to the eye that can result from RF exposure?
A. The strong magnetic fields can cause blurred vision
B. The strong magnetic fields can cause polarization lens
C. It can cause heating, which can result in the formation of cataracts
D. It can cause heating, which can result in astigmatism

T0E09 (C)
Which of the following effects on the human body are a result of exposure to high levels of RF energy?
A. Very rapid hair growth
B. Very rapid growth of fingernails and toenails
C. Possible heating of body tissue
D. High levels of RF energy have no known effect on the human body

T0E10 (D)
Why should you not stand within reach of any transmitting antenna when it is being fed with 1500 watts of RF energy?
A. It could result in the loss of the ability to move muscles
B. Your body would reflect the RF energy back to its source
C. It could cause cooling of body tissue
D. You could accidentally touch the antenna and be injured

T0E11 (B)
What is one effect of RF non-ionizing radiation on the human body?
A. Cooling of body tissues
B. Heating of body tissues
C. Rapid dehydration
D. Sudden hair loss

T0F Routine station evaluation.

T0F01 (D)
Is it necessary for you to perform mathematical calculations of the RF radiation exposure if your VHF station delivers more than 50 watts peak envelope power (PEP) to the antenna?
A. Yes, calculations are always required to ensure greatest accuracy
B. Calculations are required if your station is located in a densely populated neighborhood

C. No, calculations may not give accurate results, so measurements are always required
D. No, there are alternate means to determine if your station meets the RF radiation exposure limits

T0F02 (A)
What is one method that Amateur Radio licensees may use to conduct a routine station evaluation to determine whether the station is within the Maximum Permissible Exposure guidelines?
A. Direct measurement of the RF fields
B. Indirect measurement of the energy density at the limit of the controlled area
C. Estimation of field strength by S-meter readings in the controlled area
D. Estimation of field strength by taking measurements using a directional coupler in the transmission line

T0F03 (A)
What document establishes mandatory procedures for evaluating compliance with RF exposure limits?
A. There are no mandatory procedures
B. OST/OET Bulletin 65
C. Part 97 of the FCC rules
D. ANSI/IEEE C95.1—1992

T0F04 (B)
Which category of transceiver is NOT excluded from the requirement to perform a routine station evaluation?
A. Hand-held transceivers
B. VHF base station transmitters that deliver more than 50 watts peak envelope power (PEP) to an antenna
C. Vehicle-mounted push-to-talk mobile radios
D. Portable transceivers with high duty cycles

T0F05 (C)
Which of the following antennas would (generally) create a stronger RF field on the ground beneath the antenna?
A. A horizontal loop at 30 meters above ground
B. A 3-element Yagi at 30 meters above ground
C. A 1/2 wave dipole antenna 5 meters above ground
D. A 3-element Quad at 30 meters above ground

T0F06 (D)
How may an amateur determine that his or her station complies with FCC RF-exposure regulations?
A. By calculation, based on FCC OET Bulletin No. 65
B. By calculation, based on computer modeling
C. By measurement, measuring the field strength using calibrated equipment
D. Any of these choices

T0F07 (B)
Below what power level at the input to the antenna are Amateur Radio operators categorically excluded from routine evaluation to predict if the RF exposure from their VHF station could be excessive?
A. 25 watts peak envelope power (PEP)
B. 50 watts peak envelope power (PEP)
C. 100 watts peak envelope power (PEP)
D. 500 watts peak envelope power (PEP)

T0F08 (B)
Above what power level is a routine RF radiation evaluation required for a VHF station?
A. 25 watts peak envelope power (PEP) measured at the antenna input
B. 50 watts peak envelope power (PEP) measured at the antenna input
C. 100 watts input power to the final amplifier stage
D. 250 watts output power from the final amplifier stage

T0F09 9(D)
What must you do with the records of a routine RF radiation exposure evaluation?
A. They must be sent to the nearest FCC field office
B. They must be sent to the Environmental Protection Agency
C. They must be attached to each Form 605 when it is sent to the FCC for processing
D. Though not required, records may prove

useful if the FCC asks for documentation to substantiate that an evaluation has been performed

T0F10 (A)
Which of the following instruments might you use to measure the RF radiation exposure levels in the vicinity of your station?
A. A calibrated field strength meter with a calibrated field strength sensor
B. A calibrated in-line wattmeter with a calibrated length of feed line
C. A calibrated RF impedance bridge
D. An amateur receiver with an S meter calibrated to National Bureau of Standards and Technology station WWV

T0F11 (A)
What effect does the antenna gain have on a routine RF exposure evaluation?
A. Antenna gain is part of the formulas used to perform calculations
B. The maximum permissible exposure (MPE) limits are directly proportional to antenna gain
C. The maximum permissible exposure (MPE) limits are the same in all locations surrounding an antenna.
D. All of these choices are correct

T0F12 (C)
As a general rule, what effect does antenna height above ground have on the RF exposure environment?
A. Power density is not related to antenna height or distance from the RF exposure environment
B. Antennas that are farther above ground produce higher maximum permissible exposures (MPE)
C. The higher the antenna the less the RF radiation exposure at ground level
D. RF radiation exposure is increased when the antenna is higher above ground

T0F13 (C)
Why does the FCC consider a hand-held transceiver to be a portable device when evaluating for RF radiation exposure?
A. Because it is generally a low-power device
B. Because it is designed to be carried close to your body
C. Because it's transmitting antenna is generally within 20 centimeters of the human body
D. All of these choices are correct

T0F14 (C)
Which of the following factors must be taken into account when using a computer program to model RF fields at your station?
A. Height above sea level at your station
B. Ionization level in the F2 region of the ionosphere
C. Ground interactions
D. The latitude and longitude of your station location

T0F15 (C)
In which of the following areas is it most difficult to accurately evaluate the effects of RF radiation exposure?
A. In the far field
B. In the cybersphere
C. In the near field
D. In the low-power field

About the Author

Michael "Mick" Chesbro is a 21st-century adventurer, author, and technologist.

Michael Chesbro is a board-certified forensic examiner, a fellow of the American College of Forensic Examiners, and a diplomat of the American Board of Forensic Examiners and the American Board of Law Enforcement Experts. He holds degrees in security management, jurisprudence, and paralegal studies and is a graduate of the Federal Law Enforcement Training Center. He is professionally certified as a protection officer and security supervisor by the International Foundation for Protection Officers and is a nationally registered Emergency Medical Technician (EMT).

A former senior counterintelligence agent with U.S. Department of Defense Special Operations, Mick is now retired and devotes his time to writing, technical research, and consulting. He also serves as the director of the Auroral Radio Research Group (Radio Reconnaissance).

Mick is an amateur radio operator (call sign: KD7KLA), a GMRS radio operator (call sign: WPRR701), and an army MARS radio operator (call sign: AAR0MR). He can be found working all of the amateur radio bands in just about any mode, from Morse code to the various sound card/digital modes to phone to satellite. He runs PACTOR and can be contacted through the WinLink 2000 Network or directly on his own PBBS. Locally, you will also find him on the GMRS frequencies, CB, or MURS.

Mick's other hobbies center around outdoor activities (camping, hiking, hunting, and wilderness survival skills). He frequently incorporates radio into these activities and can often be found in the backwoods, on a mountaintop, or in some other remote location, where he can be heard operating his HAM radios.